AP® BIOLOGY
PREMIUM PREP

28th Edition

The Staff of The Princeton Review

PrincetonReview.com

The Princeton Review
110 East 42nd Street, 7th Floor
New York, NY 10017
princetonreview.com
penguinrandomhouse.com

Copyright © 2025 by TPR Education IP Holdings, LLC.
All rights reserved.

Published in the United States by Penguin Random House LLC,
New York.

Terms of Service: The Princeton Review Online Companion Tools ("Student Tools") for retail books are available for only the two most recent editions of that book. Student Tools may be activated only twice per eligible book purchased for two consecutive 12-month periods, for a total of 24 months of access. Activation of Student Tools more than twice per book is in direct violation of these Terms of Service and may result in discontinuation of access to Student Tools Services.

Please note that no part of this book may be used or reproduced in any manner for the purpose of training artificial intelligence technologies or systems.

ISBN: 978-0-593-51839-7
eBook: 978-0-593-51840-3
ISSN: 2688-1810

AP is a trademark registered by the College Board, which is not affiliated with, and does not endorse, this product.

The Princeton Review is not affiliated with Princeton University.

Editor: Laura Rose
Production Editors: Lyssa Mandel and Emma A. Parker
Production Artist: Lisa Barham
Content Contributor: Gabrielle Peterson

Manufactured in the United States of America

10 9 8 7 6 5 4 3 2 1

28th Edition

EU Contact:
Penguin Random House Ireland
32 Nassau Street
Dublin D02 YH68
https://eu-contact.penguin.ie

The Princeton Review Publishing Team
Rob Franek, Editor-in-Chief
David Soto, Senior Director, Data Operations
Stephen Koch, Senior Manager, Data Operations
Deborah Weber, Director of Production
Jason Ullmeyer, Production Design Manager
Jennifer Chapman, Senior Production Artist
Selena Coppock, Director of Editorial
Aaron Riccio, Director, Editorial Admissions Content
Orion McBean, Senior Editor
Meave Shelton, Senior Editor
Chris Chimera, Editor
Patricia Murphy, Editor
Laura Rose, Editor
Isabelle Appleton, Editorial Assistant

Penguin Random House Publishing Team
Tom Russell, VP, Publisher
Alison Stoltzfus, Senior Director, Publishing
Emily Hoffman, Managing Editor
Mary Ellen Owens, Assistant Director of Production
Suzanne Lee, Designer
Eugenia Lo, Publishing Assistant

For customer service, please contact **editorialsupport@review.com**, and be sure to include:

- full title of the book
- ISBN
- page number

Acknowledgments

The Princeton Review would like to thank Gabrielle Peterson for her valuable contribution to the 28th edition of this book.

Our gratitude also goes out to Lisa Barham, Lyssa Mandel, and Emma A. Parker for their careful attention to each page.

Contents

Get More (Free) Content viii

Part I: Using This Book to Improve Your AP Score 1
Preview: Your Knowledge, Your Expectations 2
Your Guide to Using This Book 2
How to Begin.. 3

Part II: Practice Test 1 7
Practice Test 1 9
Practice Test 1: Diagnostic Answer Key and Explanations 41
How to Score Practice Test 1 58

Part III: About the AP Biology Exam 59
The Structure of the AP Biology Exam 60
The AP Biology Exam Is Hybrid Digital 60
How the AP Biology Exam Is Scored....................................... 61
Overview of Content Topics 62
Breakdown of Free-Response Questions 64
How AP Exams Are Used.. 66
Other Resources... 67
Designing Your Study Plan 68

Part IV: Test-Taking Strategies for the AP Biology Exam 69
Preview... 70
1   How to Approach Multiple-Choice Questions 71
2   How to Approach Free-Response Questions 83
3   Using Time Effectively to Maximize Points 89

Part V: Content Review for the AP Biology Exam 93
4   Chemistry of Life 95
Elements 96
Subatomic Particles 96
Compounds... 97
Water: The Versatile Molecule... 97
Acids and Bases... 99
Organic Molecules 100
Chapter 4 Key Terms 114
Chapter 4 Drill... 115
Chapter 4 Summary... 118

5 Cell Structure and Function 119

Living Things 120

Types of Cells and Organelles 121

Transport: Traffic Across Membranes 128

Chapter 5 Key Terms 134

Chapter 5 Drill 135

Chapter 5 Summary 138

6 Cellular Energetics 139

Bioenergetics 140

Enzymes 142

Reaction Coupling and ATP 146

Photosynthesis 147

The Light Reactions 150

Cellular Respiration 153

Chapter 6 Key Terms 160

Chapter 6 Drill 161

Chapter 6 Summary 164

7 Cell Communication and Cell Cycle 165

Cell Communication 166

Feedback 168

The Cell Cycle 170

Interphase: The Growing Phase 170

Mitosis: The Dance of the Chromosomes 173

Chapter 7 Key Terms 177

Chapter 7 Drill 178

Chapter 7 Summary 180

8 Heredity 181

Haploids Versus Diploids 182

Gregor Mendel: The Father of Genetics 183

Non-Mendelian Genetics 189

Sex-Linked Traits 191

Environmental Effects on Traits 193

An Overview of Meiosis 194

Chapter 8 Key Terms 200

Chapter 8 Drill 201

Chapter 8 Summary 204

9 Gene Expression and Regulation ... 207
DNA: The Blueprint of Life ... 208
Genome Structure ... 211
DNA Replication ... 211
RNA ... 213
Translation ... 217
Regulation of Gene Expression and Cell Specialization ... 220
Mutations ... 223
Biotechnology ... 226
Chapter 9 Key Terms ... 228
Chapter 9 Drill ... 229
Chapter 9 Summary ... 232

10 Natural Selection ... 235
Natural Selection ... 236
Common Ancestry ... 239
Genetic Variability ... 240
Causes of Evolution ... 242
Species ... 244
Population Genetics ... 245
Origins of Life on Earth ... 247
Chapter 10 Key Terms ... 249
Chapter 10 Drill ... 250
Chapter 10 Summary ... 252

11 Ecology ... 255
Interacting with the Environment ... 256
Behavior ... 256
How Animals Communicate ... 258
Plant Behavior ... 260
Ecology ... 262
Population Ecology ... 267
Ecological Succession ... 269
Human Impact on the Environment ... 270
Chapter 11 Key Terms ... 272
Chapter 11 Drill ... 273
Chapter 11 Summary ... 276

12 Quantitative Skills and Biostatistics 279

Summarizing and Presenting Data 280

Types of Data 281

Types of Experiments or Questions 286

Probability 290

Hypothesis Testing.. 291

Chapter 12 Key Terms 296

Chapter 12 Drill 297

Chapter 12 Summary 299

13 Sample Free-Response Questions 301

Section II: Free Response 302

Scoring Guidelines 304

Sample Essay 306

How to Use the Sample Free-Response Questions.......... 308

Chapter 13 Summary 309

14 Laboratory 311

Lab 1: Artificial Selection... 312

Lab 2: Mathematical Modeling: Hardy-Weinberg.......... 313

Lab 3: Comparing DNA Sequences to Understand Evolutionary

Relationships with BLAST 313

Lab 4: Diffusion and Osmosis 314

Lab 5: Photosynthesis.. 314

Lab 6: Cellular Respiration 315

Lab 7: Mitosis and Meiosis 315

Lab 8: Biotechnology: Bacterial Transformation............ 316

Lab 9: Biotechnology: Restriction Enzyme Analysis of DNA............ 317

Lab 10: Energy Dynamics... 318

Lab 11: Transpiration... 318

Lab 12: Fruit Fly Behavior.. 318

Lab 13: Enzyme Activity 319

15 Chapter Drill Answers and Explanations 321

AP Biology Equations and Formulas.............................. 333

Part VI: Additional Practice Tests 335

Practice Test 2 337

Practice Test 2: Answers and Explanations 359

How to Score Practice Test 2.. 373

Practice Test 3 375

Practice Test 3: Answers and Explanations 399

How to Score Practice Test 3.. 413

Bubble Sheets 415

Online Practice Tests

Practice Test 4 .... online

Practice Test 4: Answers and Explanations .... online

Practice Test 5 .... online

Practice Test 5: Answers and Explanations .... online

Practice Test 6 .... online

Practice Test 6: Answers and Explanations .... online

at **PrincetonReview.com/prep**

As easy as **1·2·3**

1 Go to PrincetonReview.com/prep or scan the **QR code** and enter the following ISBN to register your book: **9780593518397**

2 Answer a few simple questions to set up an exclusive Princeton Review account. *(If you already have one, you can just log in.)*

3 Enjoy access to your **FREE** content!

Access Your Online Test Practice

Your AP Prep book comes with new interactive practice exams to help you prepare for digital test-taking! Find these tests in your online Student Tools, provided in two different formats:

- Digital versions with a timer option to simulate the hybrid exam experience
- Downloadable interactive PDFs with digital features like clicking your answer and typing your free-responses

Check back often as we continue to update the included AP Student Tools.

PLUS, IN YOUR ACCOUNT YOU CAN:

- Get valuable advice about the college application process, tips for essay writing, and financial aid info
- Check to see if there have been any updates or corrections to this edition
- Access comprehensive study guides, AP Biology equation and formula tables, printable Free-Response booklets, and digital flashcards to review core content

Need to report a potential **content** issue?

Contact **EditorialSupport@review.com** and include:
- full title of the book
- ISBN
- page number

Need to report a **technical** issue?

Contact **TPRStudentTech@review.com** and provide:
- your full name
- email address used to register the book
- full book title and ISBN
- Operating system (Mac/PC) and browser (Chrome, Firefox, Safari, etc.)

Look For These Icons Throughout The Book

 ONLINE PRACTICE TESTS

 PROVEN TECHNIQUES

 APPLIED STRATEGIES

 OTHER REFERENCES

Learn From the Best

The Princeton Review

With over 40 years of experience, The Princeton Review can help you achieve maximum results in minimal time.

Check out our complete line of books.

From test prep to financial aid, we've got you covered:

- ACT®, AP®, and SAT® prep—if it's on the test, it's in our books
- High school academics—put your learning on the fast track
- Admissions and financial aid guides—find and fund your perfect college match

...and much more!

PrincetonReviewBooks.com | 1-800-2-REVIEW

©2025 TPR Education IP Holdings, LLC. All Rights Reserved. The Princeton Review is not affiliated with Princeton University. ACT® is a registered trademark of ACT, Inc. AP® and SAT® are trademarks registered by the College Board, which is not affiliated with, and does not endorse, these products.

Part I
Using This Book to Improve Your AP Score

- Preview: Your Knowledge, Your Expectations
- Your Guide to Using This Book
- How to Begin

PREVIEW: YOUR KNOWLEDGE, YOUR EXPECTATIONS

Your route to a high score on the AP Biology Exam depends a lot on how you plan to use this book. Start thinking about your plan by responding to the following questions.

1. Rate your level of confidence about your knowledge of the content tested by the AP Biology Exam.

    A. Very confident—I know it all
    B. I'm pretty confident, but there are topics for which I could use help
    C. Not confident—I need quite a bit of support
    D. I'm not sure

2. If you have a goal score in mind, circle your goal score for the AP Biology Exam.

    5   4   3   2   1   I'm not sure yet

3. What do you expect to learn from this book? Circle all that apply to you.

    A. A general overview of the test and what to expect
    B. Strategies for how to approach the test
    C. The content tested by this exam
    D. I'm not sure yet

YOUR GUIDE TO USING THIS BOOK

This book is organized to provide as much—or as little—support as you need, so you can use this book in whatever way will be most helpful to improving your score on the AP Biology Exam.

*   The remainder of **Part I** will provide guidance on how to use this book and help you determine your strengths and weaknesses.

*   **Part II** of this book contains your first practice test, a Diagnostic Answer Key, detailed answers and explanations for each question, and a scoring guide. (Bubble sheets can be found at the back of this book or printed from your online Student Tools.) This is where you should begin your test preparation in order to realistically determine:
    o   your starting point right now
    o   which question types you're ready for and which you might need to practice
    o   which content topics you are familiar with and which you will want to carefully review

    Once you have nailed down your strengths and weaknesses with regard to this exam, you can focus your test preparation, build a study plan, and be efficient with your time. Our Diagnostic Answer Key will assist you with this process.

- **Part III** of this book will:
  - provide information about the structure, scoring, and content of the AP Biology Exam
  - help you to make a study plan
  - point you toward additional resources

- **Part IV** of this book will explore:
  - how to handle multiple-choice questions
  - how to write high-scoring free-response answers
  - how to manage your time to maximize the number of points available to you

- **Part V** of this book covers the content you need for your exam.

- **Part VI** of this book contains Practice Tests 2 and 3, and their answers and explanations. Compare your progress between these tests and with Practice Test 1. If you answer a certain type of question wrong several times, you probably need to review it. If you answer it wrong only once, you may have run out of time or been distracted. In either case, comparing your progress will allow you to focus on the factors that caused the discrepancy in scores and to be as prepared as possible on the day of the test.

- **Online Resources** contain three additional practice tests. Register your book to access these online resources, aka your Student Tools. Follow the study guide found there based on the amount of time you have to study for the exam.

You may choose to prioritize some parts of this book over others, or you may work through the entire book. Your approach will depend on your needs and how much time you have. Let's examine how to make this determination.

HOW TO BEGIN

1. **Take Practice Test 1**

Before you can decide how to use this book, you need to take a practice test. Doing so will give you insight into your strengths and weaknesses, and the test will also help you make an effective study plan. If you're feeling test-phobic, remind yourself that a practice test is just a tool for diagnosing yourself—it's not how well you do that matters. As long as you try your best, you can glean invaluable information from your performance to guide your preparation.

So, before you read further, take Practice Test 1 starting on page 9 of this book. Be sure to finish in one sitting, following the instructions that appear before the test.

The content in this book accurately describes what the College Board has planned for its May exam, but in the event there are any late-breaking changes, please register your book online (it's free!) to see if there are any updates.

Register your book, then go online to access three additional AP Biology practice exams, 4, 5, and 6.

AP Exams Go Digital
As of May 2025, some AP exams are fully digital and others are hybrid digital. For the latest info regarding specific AP subjects and testing accommodations, visit apcentral.collegeboard.org/exam-administration-ordering-scores/digital-ap-exams.

2. Check Your Answers

Using the Diagnostic Answer Key on page 41, follow our three-step process to identify your strengths and weaknesses with regard to the tested topics. This will help you determine which content review chapters to prioritize when studying this book. Don't worry about the explanations for now, and don't worry about missed questions. We'll get to that soon.

3. Reflect on the Test

After you take your first test, respond to the following questions:

- How much time did you spend on the multiple-choice questions?
- How much time did you spend on each free-response question?
- How many multiple-choice questions did you miss?
- Do you feel you had the knowledge to address the subject matter of the free-response questions?
- Do you feel you wrote well-organized, thoughtful responses to the free-response questions?

4. Read Part III of this Book and Complete the Self-Evaluation

As discussed in the Guide section above, Part III will provide information on how the test is structured and scored. It will also set out areas of content that are tested.

As you read Part III, re-evaluate your answers to the questions above. At the end of Part III, you will revisit and refine those questions. You will then be able to make a study plan, based on your needs and time available, that will allow you to use this book most effectively.

5. Engage with Parts IV and V as Needed

Notice the word *engage*. You'll get more out of this book if you use it intentionally than if you read it passively, hoping for an improved score through osmosis.

The strategy chapters in Part IV will help you think about your approach to the question types on this exam. Part IV will open with a reminder to think about how you approach questions now, and then close with a reflection section asking you to think about how or whether you will change your approach in the future.

The content chapters in Part V are designed to provide a review of the content tested on the AP Biology Exam, including the level of detail you need to know and how the content is tested. You will have the opportunity to assess your understanding of the content of each chapter through test-appropriate questions.

> This was just said on the previous page, but it's worth repeating! The strategy, content, and practice test chapters in this edition are up-to-date with the most recent AP Biology redesigns. If you register your book online, you'll have free access to any further updates.

6. **Take Another Test and Assess Your Performance**

   Once you feel you have developed the strategies you need and gained the knowledge you lacked, you should take Practice Test 2, which starts on page 337 of this book. You should finish in one sitting, following the instructions at the beginning of the test.

   When you are done, check your answers to the multiple-choice sections. See if a teacher will read your free-response answers and provide feedback.

   Once you have taken the test, reflect on what areas you still need to work on and revisit the chapters in this book that address those deficiencies. Once you feel confident, take Practice Test 3 and repeat the process. For extra practice, you can take Practice Tests 4, 5, and 6, which are online in your Student Tools. Through this type of reflection and engagement, you will continue to improve.

7. **Keep Working**

   After you have revisited certain chapters in this book, continue the process of testing, reflection, and engaging with the content in this book and online. Consider what additional work you need to do and how you will change your strategic approach to different parts of the test.

   As we will discuss in Part III, there are other resources available to you, including a wealth of information at AP Students, the official site of the AP Exams. You can continue to explore areas where you need more improvement and engage in those areas right up to the day of the test.

Go Online!
Check us out on YouTube for test-taking tips and techniques to help you ace your next exam at youtube.com/ThePrincetonReview.

Part II
Practice Test 1

- Practice Test 1
- Practice Test 1: Diagnostic Answer Key and Explanations
- How to Score Practice Test 1

Practice Test 1

Section I

The Exam

AP® Biology Exam

SECTION I: Multiple-Choice Questions

At a Glance

Total Time
1 hour and 30 minutes
Number of Questions
60
Percent of Total Grade
50%

DISCLAIMER: The official AP Biology Exam will be administered as a hybrid digital exam. Instructions for the digital exam may differ from this practice test.

Instructions

Section I of this examination contains 60 multiple-choice questions.

Indicate all of your answers to the multiple-choice questions on the answer sheet. Give only one answer to each question.

Use your time effectively, working as quickly as you can without losing accuracy. Do not spend too much time on any one question. Go on to other questions and come back to the ones you have not answered if you have time. It is not expected that everyone will know the answers to all the multiple-choice questions.

Your total score on the multiple-choice section is based only on the number of questions answered correctly. Points are not deducted for incorrect answers or unanswered questions.

Section I

AP BIOLOGY
SECTION I
60 Questions
Time—90 minutes

Directions: Each of the questions or incomplete statements below is followed by four suggested answers or completions. Select the one that is best in each case and then fill in the corresponding oval on the answer sheet.

Questions 1 through 5 refer to the following.

The following table lists the intracellular osmolality concentrations of four osmoconforming organisms that are known to mimic the osmotic conditions of their surroundings.

Table 1. Osmolality concentrations
of osmoconforming organisms

Organism 1	100 mOsm kg^{-1}
Organism 2	400 mOsm kg^{-1}
Organism 3	500 mOsm kg^{-1}
Organism 4	150 mOsm kg^{-1}
Organism 5	350 mOsm kg^{-1}

---

1 🔖 Mark for Review

If a large amount of salt was added to the sealed tank where organism 4 was being kept, what would be the effect on the intracellular osmolality of organism 4?

(A) Increase initially and then return to the initial state

(B) Increase and maintain the increased state

(C) Decrease and then return to the initial state

(D) Decrease and maintain the decreased state

---

2 🔖 Mark for Review

Which of the following would likely have the largest direct effect on data in the table?

(A) A mutation in a cytoskeleton protein

(B) A mutation in a protein in the sodium-potassium pump complex

(C) A mutation in a lipid in the outer chloroplast membrane

(D) A mutation in a nuclear pore channel protein

---

3 🔖 Mark for Review

Which could decrease the osmolalities of the organisms?

(A) Sodium is added to the sealed tank.

(B) Chloride ions are injected into the organism.

(C) Glucose is ingested by the organism.

(D) Water is added to the sealed tank.

GO ON TO THE NEXT PAGE.

II. Practice Test 1 | **11**

Section I

4 ☐ Mark for Review

Which of the following statements best describes the insides of these organisms compared to the environment in the tank when they reach osmoconformity?

(A) Due to the influx of water into the organisms' cells, the organisms' insides are hypertonic when compared to the tank.

(B) Compared to the tank, the loss of water out of the organisms' cells makes the organisms' insides hypotonic.

(C) The organisms' insides are isotonic since the intracellular osmolality concentrations match the osmolality concentrations of the tank.

(D) The flow of water into and out of the organisms' insides causes the cells to be retrotonic.

5 ☐ Mark for Review

A population of organisms that are not osmoconformers is dropped into the tank with Organism 1. The intracellular osmolality of Organism 1 begins to slightly decrease. Which of the following could explain this?

(A) The new organisms were hypotonic compared to the tank, and they lost water via osmosis.

(B) The new organisms expelled waste products that increased the tank's osmolality.

(C) The new organism competed with organism 1 for food and caused it to become active.

(D) The new organism ingested large amounts of fluid, and the tank fluid level decreased.

GO ON TO THE NEXT PAGE.

For more free content, visit PrincetonReview.com

Questions 6 through 9 refer to the following.

The active site of the enzyme ritzolinine (RZN45) contains three positively charged lysine residues. When ascorbic acid is present, binding JB-76, the substrate of RZN45, decreases. The reaction rate is affected by the presence of ascorbic acid as shown in the figure below. It is thought that a daily supplement of vitamin C might aid those suffering from Ritzolierre's disease, which is caused by elevated levels of RZN45.

Figure 1. Reaction rate based on ascorbic acid

---

6 Mark for Review

Which of the following is likely true?

(A) The active site of RZN45 has similarly-charged structures as JB-76.

(B) Ritzolinine is stabilized in the presence of Vitamin C.

(C) Ascorbic acid and JB-76 each have negatively charged structures.

(D) JB-76 and ascorbic acid have an identical number of amino acids.

---

7 Mark for Review

Which is the best explanation for the differing effects on reaction rate caused by ascorbic acid at low and high concentrations of JB-76?

(A) At high concentrations of JB-76, there are fewer free active sites for the ascorbic acid to bind to.

(B) At low concentrations of JB-76, there are fewer free allosteric sites for the RZN45 to bind to.

(C) At high concentrations of JB-76, there are more free allosteric sites for the ascorbic acid to bind to.

(D) At low concentrations of JB-76, there are more free active sites for the RZN45 to bind to.

GO ON TO THE NEXT PAGE.

Section I

8 — Mark for Review

A patient with Ritzolierre's disease would likely benefit from which of the following:

  I.   Injections of RZN45
  II. Injections of JB-76
  III. Injections of ascorbic acid

(A) I only

(B) II only

(C) III only

(D) II and III

9 — Mark for Review

If Figure 1 were to include additional data for higher concentrations of JB-76, how would the reaction rate change?

(A) The reaction rate would be lower than any reaction rate shown in Figure 1.

(B) The reaction rate would be higher than any reaction rate shown in Figure 1.

(C) The reaction rate would be equal to the highest reaction rate shown in Figure 1.

(D) The reaction rate would increase gradually for each concentration of JB-76 added.

10 — Mark for Review

The Galápagos Islands contain about 15 species of finches that vary in terms of beak size and strength. The differences are believed to have occurred as a result of allopatric speciation acting on the finch populations. Which of the following best explains how the finch populations developed different beaks?

(A) As finches migrated between islands, alleles were transferred between populations through the process of gene flow.

(B) Genetic drift occurred due to a hurricane near the islands, causing a bottleneck effect.

(C) Through convergent evolution, the finches developed comparable features due to being exposed to comparable selective pressures.

(D) Geographic isolation between the different islands kept the populations from breeding with each other, and each population evolved separately.

11 — Mark for Review

A cell experiences a nondisjunction event in Meiosis I. Calculate the chances of each resulting gamete exhibiting a normal karyotype.

(A) 0%

(B) 25%

(C) 50%

(D) 100%

GO ON TO THE NEXT PAGE.

Questions 12 through 16 refer to the following.

The lac operon is a set of genes that code for proteins responsible for breaking down lactose, a sugar molecule found in dairy products. A map of the lac operon is shown in Figure 1 below. When lactose is present, the catabolite activator protein (CAP) binds to the CAP site to cause release of a lac operon repressor protein that typically attaches to the Operator O region. RNA polymerase is then able to access the Promoter O region and begin transcription of the three lac operon genes, lacZ, lacY, and lacA. Figure 2 shows the relative levels of each gene's transcript as measured via RT-qPCR when CAP and the lac repressor protein are overexpressed or unexpressed in the cell.

Figure 1. Map of the lac operon

Figure 2. Relative amounts of lac operon transcript

---

12 ☐ Mark for Review

Scientists identified a mutant allele of the lac operon caused by a deletion of the Promoter P region. What is the likely effect of this mutation?

(A) No transcription and no translation of the lac operon will occur.

(B) Transcription of the lac operon will not occur, but translation will be unaffected.

(C) Transcription of the lac operon will occur, but no translation will occur.

(D) Both transcription and translation of the lac operon will be unaffected.

13 ☐ Mark for Review

Which of the following would likely produce the highest lac operon transcript levels?

(A) Overexpression of CAP and overexpression of the lac repressor protein

(B) Overexpression of CAP and unexpression of the lac repressor protein

(C) Unexpression of CAP and overexpression of the lac repressor protein

(D) Unexpression of CAP and unexpression of the lac repressor protein

GO ON TO THE NEXT PAGE.

Section I

14 ☐ Mark for Review

Based on the data, which of the following most likely describes how the binding of proteins affects the production of lac operon transcript?

(A) Because lac operon transcript levels change when certain proteins are overexpressed or unexpressed in the cell, the lac operon must interfere with the other proteins.

(B) Because lac operon transcript levels stay the same whether CAP is overexpressed or unexpressed, CAP must inhibit the production of lac operon transcript.

(C) Because lac operon transcript levels increase when CAP is unexpressed in the cell, CAP must activate the transcription of these genes.

(D) Because lac operon transcript levels decrease when CAP is unexpressed in the cell, CAP must activate the transcription of these genes.

15 ☐ Mark for Review

An additional protein was identified and found to bind to and deactivate the lac operon repressor. Based on this information, which of the following best describes why these proteins are able to bind together?

(A) Complementary nucleotides sequences

(B) Amino acid pockets with complementary conformations

(C) Opposing regions rich in cytosines and thymines

(D) Hydrophobic lipids and hydrophobic molecules

16 ☐ Mark for Review

How would the overexpression of the additional protein affect lac operon transcript levels?

(A) Lac operon transcript levels would increase.

(B) Lac operon transcript levels would decrease.

(C) Lac operon transcript levels would be unaffected.

(D) Lac operon transcript would completely disappear.

17 ☐ Mark for Review

Water molecules experience intermolecular forces when interacting with each other. Which of the following would most likely occur if water became a nonpolar molecule?

(A) Polar solutes would dissolve in water forming aqueous solutions of various concentrations.

(B) The amount of liquid water on the planet would increase, and the amount of water vapor in the air would decrease.

(C) The melting and boiling points of water would increase.

(D) Solid water would be more dense than liquid water, so ice would no longer float.

GO ON TO THE NEXT PAGE.

16 | For more free content, visit PrincetonReview.com

Questions 18 through 21 refer to the following.

A chain of three small islands was found to be the home to a small species of mouse. The angle of jaw opening was found to vary significantly. The average size angle of the maximal jaw opening found in mice at 10 locations on the three small islands is shown in Figure 1, below.

Figure 1. Angles of maximal jaw opening (in degrees) for mice found at various island positions

18 Mark for Review

A scientist wants to analyze the data to determine what factor accounts for the variation in maximal jaw angle. Which of the following analyses would best allow him to determine this cause?

(A) Plot the wind speeds on each island versus the jaw angle.

(B) Plot the heights of trees on each island versus the jaw angle.

(C) Plot the sizes of seeds on each island versus the jaw angle.

(D) Plot the differences in altitude on each island versus the jaw angle.

19 Mark for Review

Which of the following statements best supports that the jaw angle on Island B changed via punctuated equilibrium?

(A) Fossil evidence has shown that over time the jaw angle on Island B slowly increased.

(B) Fossil evidence has shown that the jaw angle of the mice on Island B increased very quickly.

(C) Fossil evidence has shown that over time the jaw angle on Island B slowly decreased.

(D) Fossil evidence has shown that the jaw angle of the mice on Island B fluctuates over time.

20 Mark for Review

Skeletons have been found that indicate that mice with jaw angles less than 30° were once found on Island B. Which explanation best accounts for this?

(A) A small jaw angle hinders the survival of mice on Island B.

(B) A large jaw angle hinders the survival of mice on Island A.

(C) A small jaw angle encourages the survival of mice on Island C.

(D) A large jaw angle encourages the survival of mice on Island A.

21 Mark for Review

If the mice from Island A and the mice from Island B were placed together, what would likely happen?

(A) The mice would mate, but it is impossible to predict jaw angle.

(B) The mice would mate, and the jaw angle would be approximately 27°.

(C) The mice would mate, and the jaw angle would be either 21° or 32°.

(D) The mice would not be capable of mating.

GO ON TO THE NEXT PAGE.

Section I

22 ☐ Mark for Review

A scientist studying osmosis placed six similarly sized pieces of potato in sodium chloride solutions of different concentrations and then measured the percent change in mass.

Table 1. % change in mass of potato samples

Sodium chloride concentration	% change in mass
0.1 M	+17.5
0.3 M	+5.0
0.5 M	−7.5
0.7 M	−16.2
0.9 M	−26.1
1.1 M	−25.1

The percent change in mass is closest to 0% at which of the following approximate concentrations of the solutions?

Ⓐ 0.01 M

Ⓑ 0.25 M

Ⓒ 0.30 M

Ⓓ 0.45 M

GO ON TO THE NEXT PAGE.

Section I

Questions 23 through 26 refer to the following.

Proteins often need the equivalent of a shipping label so that they can be sent intracellularly or extracellularly to the correct location. These labels are typically found in three forms: signal sequences, target signals, or localization signals. A signal sequence is a stretch of hydrophobic amino acid residue that is only located within polypeptides destined to fold into EITHER extracellular proteins, intermembrane proteins that will anchor to the cell surface, or proteins that will be members of the secretory pathway that packs and ships things towards the cell surface. A targeting signal is a sequence that identifies the proteins of the secretory pathway. A localization signal is a sequence that labels proteins that are destined to go to specific organelles (that are not part of the secretory pathway). The table below shows seven eukaryotic proteins, and the sequences have been identified in each.

Table 1. Sequences identified within eukaryotic proteins

Protein	Signal Sequence	Targeting Signal	Localization Sequence
HRIET1	X	X	
HAZL2			
NUH8			X
K8TE	X		
TELEE	X	X	
TMSDG	X	X	
LNACT			X

---

23 ◻ Mark for Review

Which of following could be a description for HRIET1?

(A) A protein found in the blood that plays a role in the immune system

(B) A protein found in the Golgi apparatus that helps tether vesicles

(C) A protein found in the nucleolus aiding in ribosome assembly

(D) A protein found in the cytosol serving as a cytoskeletal anchor

---

24 ◻ Mark for Review

One of the proteins was found to be a receptor for a large protein that is produced in the brain and travels through the blood until it can dock at the target cells that it stimulates. Which of the proteins is likely this receptor?

(A) TMSDG

(B) NUH8

(C) HAZL2

(D) K8TE

GO ON TO THE NEXT PAGE.

II. Practice Test 1 | 19

Section I

25 ☐ Mark for Review

Two proteins were identified in the human electron transport chain. One was found to be active at the beginning of the chain, and the other was found to be active at the final stages of the chain. Which two proteins were these?

(A) HRIET1 and HAZL2

(B) NUH8 and LNACT

(C) TELEE and K8TE

(D) TMSDG and TELEE

26 ☐ Mark for Review

If a scientist was hoping to isolate and purify HAZL2, which of the following techniques would they use?

(A) Collect only the nuclear fraction

(B) Collect only the cytosolic fraction

(C) Collect only the extracellular fraction

(D) Collect only the lipid membrane fraction

GO ON TO THE NEXT PAGE.

Questions 27 through 31 refer to the following.

In a heavily populated suburb, two cougars were once spotted roaming in a small field. Local wildlife experts, though not surprised, warned the public to be aware of their surroundings and to keep small pets protected. A group of local junior high school students were curious about the population of cougars in the area since no one they asked had ever seen one in the area. With the help of local wildlife enthusiasts and carefully placed motion-activated wildlife cameras, the group of students recorded sightings of animals in a local forest preserve for their entire four years of high school. The results are shown below.

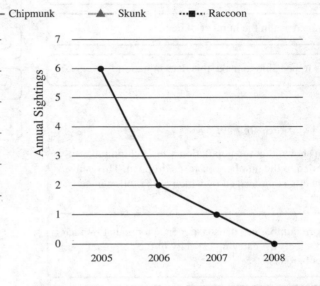

27 Mark for Review

Raccoons eat a diet rich in berries. Which statement is the most likely to describe the berry population?

(A) A hard frost eliminated nearly all of the berries in 2007, but 2008 was a milder winter.

(B) A virus was introduced in 2006 that eliminated a fly using the berry bushes as a niche habitat.

(C) Cougars destroyed much of the berry habitat in 2005.

(D) The berry population remained steady with a slight decrease due to lower rainfall in 2008.

28 Mark for Review

Although the data is limited, which of the following populations is the most likely to be a primary food source for the cougars?

(A) Skunk

(B) Deer

(C) Raccoon

(D) Chipmunk

GO ON TO THE NEXT PAGE.

Section I

29 ▢ Mark for Review

Which change will cause fewer skunks to be sighted in the future?

(A) An increase in the number of cougars

(B) An increase in the number of raccoons

(C) An increase in the number of deer

(D) It is impossible to make this determination.

30 ▢ Mark for Review

It has been shown that the population size of chipmunks is directly tied to the number of acorns dropped. This can be scientifically summarized by which of the following statements?

(A) Acorn number is a density-dependent population factor and affects the carrying capacity of the chipmunk population.

(B) Acorn number is a density-dependent population factor and does not affect the carrying capacity of the chipmunk population

(C) Acorn number is a density-independent population factor and affects the carrying capacity of the chipmunk population.

(D) Acorn number is a density-independent population factor and does not affect the carrying capacity of the chipmunk population.

31 ▢ Mark for Review

Which of the following populations does NOT demonstrate exponential growth?

I. Cougar
II. Chipmunk
III. Deer

(A) I only

(B) I and II

(C) I and III

(D) I, II, and III

GO ON TO THE NEXT PAGE.

Section I

Questions 32 through 35 refer to the following.

A certain gene has been identified on chromosome 2 in a species of butterfly. The mRNA transcript has been found to be 1,578 base pairs in length. A portion of the transcript is shown below. The area shown in grey is part of a known ribosome binding site.

Figure 1. Portion of the mRNA transcript of a species of butterfly

5' UUACGUGCAUGCGUAGCUAGCUAGCUAUGCUAUCGCUUUAAGAGCAAAAAA 3'
15   30   45

---

32 ☐ Mark for Review

A scientist wants to determine the specific effect of deleting bases 0–15. Which of the following should she use as her dependent variable?

(A) Levels of transcription

(B) Levels of translation

(C) Levels of protein synthesis

(D) There will be no effect of deleting these bases.

---

33 ☐ Mark for Review

A transcript was identified in the nucleus that is similar to this transcript. How is it likely different from the mRNA transcript?

(A) The nuclear transcript likely contained introns.

(B) The nuclear transcript likely contained exons.

(C) The nuclear transcript likely contained a poly-A tail.

(D) The nuclear transcript likely contained a 3' GTP-Cap.

---

34 ☐ Mark for Review

RNA polymerase transcribes DNA into mRNA by matching base pairs of nucleotides. Which of the following would be the segment of DNA on the coding strand corresponding to bases 0–6 in Figure 1?

(A) 5' AAUGCA 3'

(B) 5' TTACGT 3'

(C) 3' AATGCA 5'

(D) 3' TTUGCT 5'

---

35 ☐ Mark for Review

Where on the strand shown in Figure 1 is a transcription factor likely to bind?

(A) Bases 0 to 15

(B) Bases 20 to 25

(C) Bases –20 to –15

(D) None of the above

---

36 ☐ Mark for Review

In a small ecosystem, it is estimated that producers equal approximately 2,987 g/m^2 of biomass. What would be a good guess of how much biomass is found within secondary consumers?

(A) 5,974 g/m^2

(B) 2,987 g/m^2

(C) 298.7 g/m^2

(D) 29.87 g/m^2

GO ON TO THE NEXT PAGE.

II. Practice Test 1 | **23**

Section I

Questions 37 through 40 refer to the following.

The enzymatic catalysis of a reaction essential in the production of dog saliva is mediated by the protein ARKKKK60491. ARKKKK60491 is a homodimer, with each subunit being 337 amino acids. Figure 1 below indicates a modular structure of the homodimer. Three positions are indicated on the model. Position A is the active site. Position B is a region known to have a large number of nonpolar residues. Position C is known to have a large number of charged residues.

Figure 1. Modular structure of the homodimer

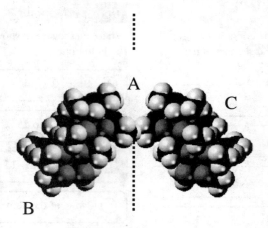

37 Mark for Review

A mutation is discovered in the gene for ARKKKK60491 that converts positively charged lysine residue into negatively charged glutamic acid residue. This single change directly impacts the location where the substrate to ARKKKK60491 attaches during catalysis. Which position is likely affected by this change?

(A) Position A

(B) Position B

(C) Position C

(D) None of the positions will be affected.

38 Mark for Review

Which of the following is Position B most likely to assist the protein with?

(A) Stabilization of the substrate into the transition state during catalysis

(B) Transportation of catalyzed cargo from the nucleus to the mitochondria

(C) Creation of a homodimer through catalysis of homodimerization

(D) Attachment to the membrane by embedding into the phospholipid bilayer

GO ON TO THE NEXT PAGE.

Section I

39 ☐ Mark for Review

Which of the following statements best predicts the effect on the protein's structure if the subunits of the homodimer fail to attach?

(A) The primary structure will change because the order of amino acids will differ.

(B) The secondary structure will differ because hydrogen bonds will occur in different locations, and the protein will fold into a different shape.

(C) The tertiary structure will be destroyed because the side chain interactions will not occur and the protein will not fold into a 3-dimensional shape.

(D) The quaternary structure will be destroyed because the different amino acid chains will not join together.

40 ☐ Mark for Review

According to Figure 1, within the protein ARKKKK60491, which of the following locations contain polypeptide N-termini?

   I.   Position A
  II.   Position B
 III.   Position C

(A) I only

(B) I and II

(C) II and III

(D) I and III

GO ON TO THE NEXT PAGE.

Questions 41 through 44 refer to the following.

There are four major types of macromolecules, and each type has specific functions and characteristics. Proteins are made of amino acids and function as enzymes, receptors, hormones, signals, and receptors. Lipids are made of fatty acids and glycerol and function as membrane components and energy storage. Carbohydrates are made of monosaccharides and function as energy storage. Nucleic acids are made of nucleotides and provide genetic information for organisms.

41 Mark for Review

Fats are a type of lipid that are generally soluble in organic solvents and generally insoluble in water. Which of the following questions will best guide the researchers in determining if a substance is a fat?

(A) At what temperature does the substance boil?

(B) What type of mixture does the substance form when mixed with water?

(C) Does treating the substance with acid affect its appearance?

(D) Have chemical reactions been recorded between the substance and other chemicals?

42 Mark for Review

Sugars are a type of carbohydrate typically made up of hydrogen, carbon, and oxygen. The hydrogen:oxygen ratio is typically 2:1. Which of the following statements explains how carbohydrates function as a stored form of energy?

(A) Excess glucose is stored as glycogen in muscles and the liver.

(B) Glucose is broken down during photosynthesis to act as an energy source.

(C) Glycerol is released into the bloodstream when stored fat is used as an energy source.

(D) Glucagon is a hormone that stores excess energy in the pancreas.

43 Mark for Review

All four types of macromolecules are organic compounds. Which of the following statements describes a difference between organic and inorganic compounds?

(A) Only organic compounds contain oxygen.

(B) Only organic compounds contain carbon atoms covalently bonded to hydrogen atoms.

(C) Only organic compounds contain hydrogen bonds.

(D) Only organic compounds contain ionizing chemical groups.

44 Mark for Review

Figure 1. Structure of adenine

Which of the following could NOT be represented by Figure 1?

(A) A monomer used to build a protein

(B) A nucleotide

(C) A reactant used to synthesize nucleic acids

(D) A chemical used to relay genetic information

GO ON TO THE NEXT PAGE.

Section I

Questions 45 through 49 refer to the following.

A nettle has been discovered, native to a nation, that produces tiny flowers that come in two colors: white and pink. The petals of the flowers come in two varieties: ruffled or slotted. After years of studying them, it has been widely acknowledged that both traits follow classic Mendelian inheritance patterns of dominance. The table below shows floral phenotypes in a small garden in an isolated greenhouse, where the dominant phenotypes are the most plentiful and researchers assume the population is in Hardy-Weinberg equilibrium.

Table 1. Population of various floral phenotypes

Phenotype	#
White	4900
Pink	400
Ruffled	700
Slotted	4600

---

45 ☐ Mark for Review

A true-breeding white nettle is crossed with a true-breeding pink nettle. What will be the ratio of the offspring?

Ⓐ 4 White : 0 Pink

Ⓑ 3 White : 1 Pink

Ⓒ 1 White : 1 Pink

Ⓓ 1 White : 3 Pink

---

46 ☐ Mark for Review

The F_1 generation, the offspring in the cross of the previous question, was crossed with a true-breeding pink nettle. What percentage of offspring would have pink flowers?

Ⓐ 0%

Ⓑ 25%

Ⓒ 50%

Ⓓ 100%

---

47 ☐ Mark for Review

A cross is performed between one nettle that is known to be heterozygous for both traits and a nettle that is pink and ruffled. Which of the following would lead researchers to believe the traits are linked?

Ⓐ Several white flowers are identified with ruffled petals.

Ⓑ Nearly all white flowers also have slotted petals, and nearly all pink flowers have ruffled petals.

Ⓒ Petals are discovered that are both ruffled and slotted.

Ⓓ Flowers are discovered that have two colors of flowers and two shapes of petals.

---

48 ☐ Mark for Review

What frequency of the alleles in the greenhouse garden are recessive for the flower color trait?

Ⓐ 0.14

Ⓑ 0.27

Ⓒ 0.38

Ⓓ 0.56

GO ON TO THE NEXT PAGE.

II. Practice Test 1 | **27**

Section I

49 ▢ Mark for Review

What is the frequency of plants in the greenhouse garden that are heterozygous for the flower color trait?

(A) 0.27

(B) 0.39

(C) 0.50

(D) 0.74

GO ON TO THE NEXT PAGE.

Questions 50 through 53 refer to the following.

G-protein coupled receptors (GPCRs) are a common type of protein receptor containing seven transmembrane segments and bearing intracellular and extracellular portions. The structures of a typical GPCR and its associated partners are shown below. GPCRs are coupled to G-proteins consisting of three subunits (α, β, and γ) that bind the nucleotides GTP or GDP on the intracellular side of the GPCR. The presence or absence of a ligand on the extracellular side of the GPCR determines whether GDP or GTP will bind to the intracellular G-protein. Extracellular ligand binding initiates GDP being exchanged for GTP and the separation of the G-protein. One subunit will travel to join with a secondary partner and will initiate a cascade of signaling effects within the cell. In the example shown below, the secondary partner is adenylyl cyclase, and the signaling is cAMP upregulation.

Figure 1. G-protein coupled receptor joining with adenylyl cyclase

50. Mark for Review

A setup like a GPCR is unnecessary in which of the following situations?

(A) The signaling event is very specific.

(B) The ligand is a hormone traveling in the bloodstream.

(C) The effector molecules are located inside the cytoplasm.

(D) The ligand is a small nonpolar molecule.

51. Mark for Review

A scientist wants to confirm the first step in the activation of cAMP-dPK by using drugs to knock out different components of the receptor. Which of the following processes should the scientist most likely target?

(A) The exchange of GTP for GDP

(B) The binding of epinephrine

(C) The separation of a-GTP

(D) Phosphorylation of intracellular enzymes

GO ON TO THE NEXT PAGE.

Section I

52 ▢ Mark for Review

Which of the following situations is most similar to extracellular binding to a GPCR?

- Ⓐ A police officer pulling over a speeding car on the expressway
- Ⓑ A celebrity chef visiting the kitchen of another restaurant
- Ⓒ A driver giving their order to a person through a drive-up window
- Ⓓ A taxi driver picking up a group of passengers

53 ▢ Mark for Review

What does a GPCR's capability of making a ligand-induced conformational change allow?

- Ⓐ Binding of the ligand to the phospholipid head groups within the membrane
- Ⓑ Communication between the outside of the cell and the inside of the cell
- Ⓒ A solid attachment to the intermembrane region of the cell membrane
- Ⓓ Intracellular binding of the GDP-associated G-protein complex

GO ON TO THE NEXT PAGE.

Section I

Questions 54 through 58 refer to the following.

A small family farm has documented carrot growth for a hundred years in a Midwest microclimate. The average lengths and masses of the carrots are shown in the table below. The same ecosystem was home to a species of vegetable weevil that attacks the carrots underground. Other than carrots, the weevils also preyed on radishes found in the ecosystem. This particular species of weevil spends its whole life beneath the soil, and the average depth at which these weevils were found each year is shown in the table below.

Table 1. Average lengths and masses of carrots and depth at which weevils are found (by decade)

Year	1920	1930	1940	1950	1960	1970	1980	1990	2000	2010	2020
Average carrot mass (g)	61	62	61	65	58	61	60	62	65	61	59
Average carrot length (cm)	21.5	21.2	20.3	19.8	19.0	18.7	18.1	17.6	17.5	17.9	17.7
Average weevil depth (cm)	19.2	20.2	20.0	19.7	19.3	20.3	19.4	20.2	20.1	19.8	19.9

---

54 ☐ Mark for Review

Which carrot phenotype was selected for over time?

(A) Thinner carrots

(B) Longer carrots

(C) Shorter carrots

(D) Heavier carrots

---

55 ☐ Mark for Review

If a weevil was introduced to the ecosystem that lived at a depth of 17 centimeters, how would the carrot length change in the coming years?

(A) It would likely return to its original long size.

(B) It would likely get longer than it has been in the past.

(C) It would likely get shorter than it has been in the past.

(D) It would not likely change at all.

---

56 ☐ Mark for Review

The average carrot consumer prefers long carrots. What should be done to promote the growth of longer carrots?

   I.   Remove the weevil population
  II.   Cross the longest carrot of each generation
 III.   Introduce a different carrot predator that lives at 10 centimeters

(A) I only

(B) I and II

(C) II and III

(D) I, II, and III

---

57 ☐ Mark for Review

If true, which of the following would account for the average weevil depth changing to 17 centimeters in 2030?

(A) The average carrot length changed to 15 centimeters.

(B) The radishes completely disappeared from the garden.

(C) Goats arrived to eat the tops of the carrot plants above the surface.

(D) A small percentage of the weevil population died from a virus.

GO ON TO THE NEXT PAGE.

II. Practice Test 1 | **31**

Section I

58 ☐ Mark for Review

According to Table 1, which of the following traits is most linked to the carrot's fitness?

(A) Weight

(B) Length

(C) Color

(D) Flavor

59 ☐ Mark for Review

A dance by the bird of paradise is one of the most highly complex mating rituals. Precision and flamboyancy are key to wooing a mate. The natural predator of the bird of paradise also enjoys the spirited dance and often strikes during the courtship ritual. Which of the following statements is likely true?

(A) Sexual selection of the most exciting dancers is the driving selective pressure.

(B) Selective pressure against the exciting dancers is the driving selective force.

(C) The bird of paradise will soon become extinct if the predatory action does not cease.

(D) There can exist a careful balance between multiple opposing selective pressures.

60 ☐ Mark for Review

Antibiotic resistance occurs when drugs or chemicals used to treat an illness become less effective. Some varieties of *Neisseria gonorrhoeae* have become resistant to the antibiotic penicillin. Which of the following best explains how the bacteria became resistant?

(A) The bacteria experienced hybrid vigor through gaining improved traits from their parents.

(B) Through natural selection, bacteria with a random mutation that provided resistance survived and multiplied to create more bacteria.

(C) The bacteria were influenced by other species in their environment through coevolution.

(D) The bacteria evolved to fill different niches from other species in the environment through adaptive radiation.

STOP

END OF SECTION I

IF YOU FINISH BEFORE TIME IS CALLED, YOU MAY CHECK YOUR WORK ON THIS SECTION. DO NOT GO ON TO SECTION II UNTIL YOU ARE TOLD TO DO SO.

Section II

AP BIOLOGY
SECTION II
6 Questions
Writing Time—90 minutes

DISCLAIMER: The official AP Biology Exam will be administered as a hybrid digital exam. Instructions for the digital exam may differ from this practice test.

Directions: Questions 1 and 2 are long free-response questions that should require about 25 minutes each to answer and are worth 8–10 points each. Questions 3 through 6 are short free-response questions that should require about 10 minutes each to answer and are worth 4 points each.

Read each question carefully and completely. Write your response in the printable free-response booklet found in your online Student Tools. Only material written in the space provided will be scored. Answers must be written out in paragraph form. Outlines, bulleted lists, or diagrams alone are not acceptable unless specifically requested.

1. Mark for Review

The microbiome is a collection of microbes, such as bacteria and viruses, that live in/on the human body. Sites of colonization include the skin, nasal passages, mouth, and gut. Links between the microbiome and many different conditions, such as allergies, obesity, depression, and many others have been suggested. During antibiotic use, the body can often suffer repercussions due to the destruction of the natural microbiome. The bacterial community composition varies greatly between body sites. Even among the skin microbiome, there is a great variety in the amounts and types of bacteria found in different locations.

    A.   Bacterial communities containing different species of bacteria are no different than other communities with multiple species. **Explain** what factors contribute to a carrying capacity.

An experiment was performed to assess the results of skin bacterial transplants to alternative sites of the body. A 2-inch square area of interdigital skin was swabbed with a sterile swab to collect bacteria. An additional swab was opened but not touched to the skin. Each swab was then rubbed on either an alternate skin site or the same type of skin site. After 7 days, the skin of the site of deposition was swabbed and plated on a dish to assess bacterial growth. The types of bacteria present were then cataloged. Results are shown in Figures 1, 2, and 3.

Figure 1. Relative composition of bacterial communities found in each of four skin microbiomes

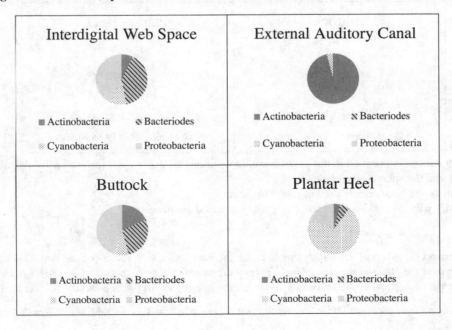

GO ON TO THE NEXT PAGE.

Figure 2. The results of rubbing a swab of interdigital skin on four skin sites

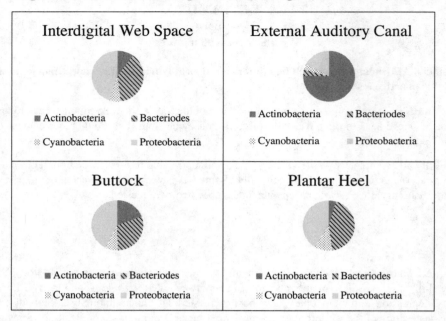

Figure 3. The results of rubbing a clean swab on four skin sites

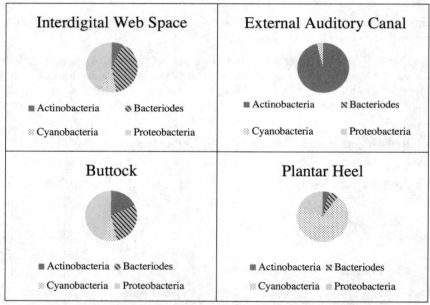

B. Two important controls were used in this experiment. In addition, the experiment includes a 7-day waiting period between when the bacterial transplants take place.
   i. **Identify** the two controls used in this experiment.
   ii. **Justify** why the two controls and 7-day waiting period are necessary.

C.
   i. **Interpret** the effect of the transplants between the interdigital web space and the buttock area.
   ii. **Interpret** the effect of the transplants between the interdigital web space and the external auditory canal area.
   iii. **Interpret** the effect of the transplants between the interdigital web space and the plantar heel area.

D. If each deposition site is pretreated with known antibiotics, **explain** how this would likely affect the results.

GO ON TO THE NEXT PAGE.

Section II

2 ☐ Mark for Review

Researchers were interested in the effect of temperature on the respiration rate of crickets.

A.

    i.   **Describe** how CO_2 is a measure of respiration.

    ii. **Explain** why temperature might affect the rate of respiration.

Three chambers were prepared and kept at three different temperatures. A cricket was placed in each chamber. The CO_2 levels in the chamber were recorded at 0 minutes, 5 minutes, and 10 minutes to assess the changing levels of CO_2 over time. This process was repeated 50 times at each temperature. The results are indicated in Table 1.

Table 1. Levels of CO_2 over time

Temperature (°C)	CO_2 levels (ppm $\pm$ 2 SE_x)		
	0 min	5 min	10 min
20	1034 ± 26	1074 ± 251	1112 ± 71
35	1560 ± 55	1670 ± 120	1785 ± 100
50	2021 ± 35	2040 ± 55	2051 ± 5

B.   On a set of axes, **construct** an appropriately labeled graph to demonstrate the CO_2 levels with changing temperature. Please include all temperatures, timepoints, and error bars.

C.   **Identify** which temperature was associated with the highest respiration rate in the crickets.

D.

    i.   **Predict** what would happen to the CO_2 levels in the chamber with an additional 5°C increase above the highest temperature.

    ii. **Justify** your prediction.

GO ON TO THE NEXT PAGE.

II. Practice Test 1 | **35**

3 🔖 Mark for Review

A researcher was trying to determine why a strain of influenza spreads more easily than other strains. She

Section II

4 ▢ Mark for Review

The school is growing tomatoes using a hydroponics system. Hydroponics is a way of growing plants in a controlled system without traditional soil containing nutrients. Instead, nutrients that would be supplied by soils are added to water tanks at specific concentrations before the solution is pumped onto the plant root systems. This type of system allows for increased control over nutrient levels and allows for reduced water usage compared to outdoor crops because of minimal evaporation.

An out-of-control thermostat in the science wing of the school resulted in a temperature that is known to increase thylakoid permeability to H^+ ions.

A. **Identify** which cellular process thylakoids are associated with.

B. **Explain** why ions in the hydroponic tank cannot pass into the roots by simple diffusion.

C. **Predict** the immediate effect this would have on the cell's ability to produce reactants necessary for carbon fixation.

D. **Justify** your prediction.

GO ON TO THE NEXT PAGE.

5 Mark for Review

In Figure 1, the values above branches indicate bootstrapped confidence values. Branch lengths are proportional to percent similarity in viral envelope gene sequence

C. In 50 years, new data have identified that TBEwest(Hypr) is no longer in circulation, but it seems to have evolved in two directions: TBEwest1 and TBEwest2. Using the empty tree below, correctly **indicate** TBEwest(neu), TBEwest(K

Section II

6 Mark for Review

In a city in Australia there is a population of moths that live in the grass, rarely taking flight except for short journeys. There are two phenotypes for wing color: green and beige. The largest predators of moths are birds and bats that prey upon the moths as they rest upon blades of grass. Green moths fare better in wet conditions when the grass takes on a lush green color, and beige moths fare better in dry conditions when the grass turns a dry brown color.

Figure 1. Percentage of moths of each color

A. **Identify** which color moth was the most plentiful between 1969 and 1995.

B. **Explain** why the lines of the graph never exceed 100.

C. **Predict** during which years there was a drought.

D. **Explain** how natural selection led to the rise of the beige moth in the late 1970s.

WHEN YOU ARE FINISHED WRITING, CHECK YOUR WORK ON SECTION II IF TIME PERMITS.

STOP

END OF EXAM

Practice Test 1: Diagnostic Answer Key and Explanations

AP Biology Premium Prep

PRACTICE TEST 1: DIAGNOSTIC ANSWER KEY

Let's take a look at how you did on Practice Test 1. Follow our three-step process in the diagnostic answer key below to identify your strengths and weaknesses with regard to the tested topics. This will help you determine which content review chapters to prioritize when studying this book.

STEP 1 » Check your answers and mark any correct answers with a ✔ in the appropriate column.

Multiple Choice							
Q #	Ans.	✔	Chapter #, Section Title	Q #	Ans.	✔	Chapter #, Section Title
1	B		**5,** Transport: Traffic Across Membranes	27	D		**11,** Ecology
2	B		**5,** Transport: Traffic Across Membranes	28	B		**11,** Ecology
3	D		**5,** Transport: Traffic Across Membranes	29	D		**11,** Ecology
4	C		**5,** Transport: Traffic Across Membranes	30	A		**11,** Population Ecology
5	A		**5,** Transport: Traffic Across Membranes	31	D		**11,** Population Ecology
6	C		**6,** Enzymes	32	B		**9,** Mutations
7	A		**6,** Enzymes	33	A		**9,** Regulation of Gene Expression & Cell Specialization
8	C		**6,** Enzymes	34	B		**9,** RNA
9	C		**6,** Enzymes	35	D		**9,** Regulation of Gene Expression & Cell Specialization
10	D		**10,** Species	36	D		**11,** Ecology
11	A		**8,** An Overview of Meiosis	37	A		**6,** Enzymes
12	A		**9,** Mutations	38	D		**6,** Enzymes
13	B		**9,** Regulation of Gene Expression & Cell Specialization	39	D		**6,** Enzymes
14	D		**9,** Regulation of Gene Expression & Cell Specialization	40	C		**6,** Enzymes
15	B		**9,** Regulation of Gene Expression & Cell Specialization	41	B		**4,** Organic Molecules
16	A		**9,** Regulation of Gene Expression & Cell Specialization	42	A		**4,** Organic Molecules
17	D		**4,** Water: The Versatile Molecule	43	B		**4,** Organic Molecules
18	C		**10,** Genetic Variability	44	A		**4,** Organic Molecules
19	B		**10,** Species	45	A		**8,** Gregor Mendel: The Father of Genetics
20	A		**10,** Genetic Variability	46	C		**8,** Gregor Mendel: The Father of Genetics
21	A		**10,** Species	47	B		**8,** Non-Mendelian Genetics
22	D		**5,** Transport: Traffic Across Membranes	48	B		**8,** Gregor Mendel: The Father of Genetics
23	B		**7,** Cell Communication	49	B		**8,** Gregor Mendel: The Father of Genetics
24	D		**7,** Cell Communication	50	D		**7,** Cell Communication
25	B		**7,** Cell Communication	51	B		**7,** Cell Communication
26	B		**5,** Types of Cells & Organelles	52	C		**7,** Cell Communication

42 | For more free content, visit PrincetonReview.com

Multiple Choice—Continued

Q #	Ans.	✔	Chapter #, Section Title	Q #	Ans.	✔	Chapter #, Section Title
53	B		7, Cell Communication	57	B		10, Causes of Evolution
54	C		10, Causes of Evolution	58	B		10, Causes of Evolution
55	C		10, Causes of Evolution	59	D		10, Causes of Evolution
56	B		10, Causes of Evolution	60	B		10, Causes of Evolution

Free Response

Q #	Ans.	✔	Chapter #, Section Title
1A	See Explanation		11, Ecology
1B	See Explanation		10, Natural Selection
1C	See Explanation		10, Natural Selection
1D	See Explanation		10, Natural Selection
2A	See Explanation		6, Cellular Energetics
2B	See Explanation		6, Cellular Energetics
2C	See Explanation		6, Cellular Energetics
2D	See Explanation		6, Cellular Energetics
3A	See Explanation		9, Gene Expression & Regulation
3B	See Explanation		9, Gene Expression & Regulation
3C	See Explanation		9, Gene Expression & Regulation
3D	See Explanation		9, Gene Expression & Regulation
4A	See Explanation		6, Cellular Energetics
4B	See Explanation		4, Chemistry of Life
4C	See Explanation		4, Chemistry of Life
4D	See Explanation		4, Chemistry of Life
5A	See Explanation		10, Natural Selection
5B	See Explanation		10, Natural Selection
5C	See Explanation		10, Natural Selection
5D	See Explanation		10, Natural Selection
6A	See Explanation		10, Natural Selection
6B	See Explanation		10, Natural Selection
6C	See Explanation		10, Natural Selection
6D	See Explanation		10, Natural Selection

AP Biology Premium Prep

 Tally your correct answers from Step 1 by chapter. For each chapter, write the number of correct answers in the appropriate box. Then, divide your correct answers by the number of total questions (which we've provided) to get your percent correct.

CHAPTER 4 TEST SELF-EVALUATION

CHAPTER 5 TEST SELF-EVALUATION

CHAPTER 6 TEST SELF-EVALUATION

CHAPTER 7 TEST SELF-EVALUATION

CHAPTER 8 TEST SELF-EVALUATION

CHAPTER 9 TEST SELF-EVALUATION

CHAPTER 10 TEST SELF-EVALUATION

CHAPTER 11 TEST SELF-EVALUATION

 Use the results above to customize your study plan. You may want to start with, or give more attention to, the chapters with the lowest percents correct.

AP Biology Premium Prep

PRACTICE TEST 1 EXPLANATIONS

Section I: Multiple Choice

1. **B** Adding salt would increase the osmolality (a similar measure to osmolarity), and since Organism 4 is an osmoconformer it would change its intracellular osmolality/osmolarity to mimic the surroundings. This means that it would increase its osmolality and maintain it unless further changes are made to the system.

2. **B** The only choice that has direct ties to maintaining osmotic balance is the sodium-potassium pump.

3. **D** Since the organisms mimic their surroundings, the only choice that would decrease their osmolalities would be one that decreases the osmolality of their surrounding tank. Choice (A) increases the tank osmolality. Choices (B) and (C) directly increase the organism osmolality. Choice (D) decreases the tank's osmolality.

4. **C** The goal of an osmoconformer is to mimic the osmolarity of its surroundings. An isotonic solution has the same concentration as its surroundings. Therefore, an osmoconformer would become isotonic to its tank.

5. **A** If Organism 1 decreases osmolality, it means that the tank must have decreased in osmolality. In other words, it gained water or lost solute. Choice (B) increases the tank solute. Choice (C) does not directly relate to the osmolality. Choice (D) decreases the water. Only (A) would cause the tank to gain water as water is removed from the new organism by osmosis.

6. **C** RZN45 is the enzyme and it binds to JB-76. Ascorbic acid (vitamin C) is an inhibitor of this reaction and binding to JB-76 is reduced, which suggests the inhibitor binds to the same place as the substrate. The graph confirms that the inhibition is competitive, since it shows a decreased reaction rate when ascorbic acid is present. Choice (A) is not true because similarly charged structures would repel each other and RZN45 and JB-76 bind to each other. There is nothing to suggest that (B) is correct. Choice (C) would make sense since both ascorbic acid and JB-76 bind to the same place, so they probably have similar structures. Acids are not made up of amino acids, so eliminate (D).

7. **A** The graph shows that at low/medium levels of substrate, the reaction rate is low in the presence of ascorbic acid and then at high levels of substrate the reaction rate is high, even with ascorbic acid present. This is because with lots of substrate, the ascorbic acid does not stand a chance at finding a free active site to bind to because the superfluous amount of substrate is likely to be bound at most of the sites.

8. **C** Ritzoierre's disease is caused by elevated enzyme levels. Only option III would decrease the enzyme activity levels. Options I and II would raise the levels and make the disease worse.

9. **C** As shown in the graph, the amount of substrate has reached a saturation point, and adding additional substrate will not increase the rate further. The reaction rate will remain at the same high level.

II. Practice Test 1: Diagnostic Answer Key and Explanations | **45**

AP Biology Premium Prep

10. **D** The islands are separated by ocean. Allopatric speciation occurs when populations are isolated from each other. The finch populations are far away from each other and are likely to mate only with those on the same island. Choice (A) implies that the finches are able to cross between islands, and (B) refers to a natural disaster. Neither is mentioned in the information provided. Choice (C) describes the finches developing similar features, but they developed different types of beaks. Only (D) references the isolated populations associated with allopatric speciation.

11. **A** If meiosis occurred as it should, there should be 4 haploid gametes, each with one long and one short chromosome. When there is a nondisjunction in Meiosis I, the two long homologous chromosomes failed to separate. This causes two gametes to get zero copies of the long chromosome and two gametes to get an extra copy. Therefore, there is a 0% chance any of the gametes will have a normal karyotype.

12. **A** The promoter region is upstream of the transcription start site. This is important for polymerase binding and initiation of transcription. Without it, there would be no transcription and subsequently there could not be any translation.

13. **B** CAP triggers transcription of the lac operon and the lac operon repressor inhibits transcription of the lac operon. Therefore, high levels of CAP and low levels of the repressor will have the highest expression of the lac operon.

14. **D** If lac operon transcript levels decrease when CAP is unexpressed, that indicates that CAP is important for activating the lac operon. If levels increased, that would indicate CAP is a repressor; eliminate (C).

15. **B** Two proteins fitting together would not occur from similarities in their nucleotide sequences, which is referenced in (A) and (C). As proteins, they should not be formed from lipids, so (D) is incorrect. Proteins are formed from amino acids, and complementary conformations would allow them to bind together.

16. **A** The new protein appears to deactivate the lac repressor protein, so the lac operon will no longer be repressed. Therefore, lac operon transcript levels should increase.

17. **D** Solvents dissolve solutes that are similar to them, so polar solutes would no longer dissolve in water if water turned nonpolar. Choice (B) is incorrect, because nonpolar molecules are more likely to be found as gases, so water vapor should increase. Less liquid water would be found on Earth due to the lack of strong bonds holding it together, which also ties into the melting and boiling points decreasing instead of increasing. The polarity of water allows it to hydrogen bond and form a less dense solid, so (D) describes what would happen if the polarity changed.

18. **C** The wind speed, tree height, and altitude do not tie directly into the function of the jaw and the size of the opening, so eliminate (A), (B), and (D). Only the size of seeds that the mice eat would directly relate to the mouth size, so plotting this data would allow the scientist to identify the relationship between the two factors.

19. **B** Punctuated equilibrium is defined as an isolated episode of rapid development of a population. A slow change, described in (A) and (C), or a fluctuation, described in (D), do not suggest a quick change.

20. **A** Since the jaw angle on island B is now large, the smaller jaw angle must have disappeared. The small jaw angle is likely selected against, and only mice with large jaws survive on Island B. The survival on the other islands is not directly relevant.

21. **A** It says in the passage that the mice are the same species. This means they should be able to mate with each other, so eliminate (D). Since we do not know anything about the genetics and inheritance of jaw angle, it is not possible to predict how the jaw angle would be affected.

22. **D** The percent change in mass will be closest to 0% when the solutions are isotonic. In other words, the potato cells should stay the same mass when they are soaked in a sodium chloride solution with a similar concentration as the cells. The percent change crosses over from a gain to a loss between $0.3\ M$ and $0.5\ M$, so the isotonic concentration must occur between these two molarities. Only (D) describes a number in between these concentrations.

23. **B** HRIET1 has a signal sequence AND a targeting signal, so it must be part of an organelle that is in the secretory pathway that helps things exit the cell. The Golgi apparatus is the only organelle among these choices that functions in this pathway.

24. **D** The receptor must be in the membrane on the surface of the cell. It should have a signal sequence, but it should not have a targeting signal or a localization sequence. The only protein that meets these requirements is K8TE.

25. **B** The electron transport chain in humans occurs in mitochondria. Proteins destined for the mitochondria should have a localization sequence. The only two proteins that have a localization sequence are NUH8 and LNACT.

26. **B** HAZL2 has none of the sequences or signals represented in the data table. Proteins without these features would remain in the cytosol where it is assembled, so a scientist would have to collect samples that contain cytosol.

27. **D** The raccoon population seems steady with a slight dip in 2008. If raccoons rely on berries as a primary food source, then it is likely that the berry population would have also remained stable.

28. **B** The cougar population is shown to be decreasing over time. One possible explanation for the decline may be due to a disappearing food source, so the food source should also show a decrease over time. The only animal that also shows a decreasing population is deer.

29. **D** There is no information about what species is a predator of skunks, and the population seems to be stable over time, so this is impossible to determine. Even though the skunks and raccoons are both stable, there is no indication that they are related to each other.

30. **A** If the chipmunk is dependent on the acorns, then the acorns affect the carrying capacity, so eliminate (B) and (D). Since the quantity of acorns would affect a large population differently than a small population, this makes it a density-dependent factor.

AP Biology Premium Prep

31. **D** None of the populations demonstrate exponential growth because none of the lines on the graph increase exponentially. All three species stay the same or decrease.

32. **B** The bases at the 5' side of the transcript are in the untranslated region because this is generally the site for ribosome binding. Without this region, the ribosome couldn't bind, and there would be no translation; eliminate (D). Transcription would not be affected, so eliminate (A). Protein synthesis would be affected, but it would be tough to know exactly how, so (C) is not the best answer.

33. **A** Before the transcripts are completed and leave the nucleus, they undergo certain modifications including the removal of introns and the addition of a 5' GTP cap and a 3' poly-A tail. The transcript in the nucleus would not have undergone these modifications yet, so it would contain introns.

34. **B** The coding strand is the strand that is identical to the mRNA transcript, except that the mRNA has the nucleotide uracil in place of the nucleotide thymine. The noncoding strand is the partner strand to the coding strand and used as a template during transcription. Therefore, the answer should match the transcript shown in the image with "T's" in place of any "U's."

35. **D** A transcription factor binds to DNA before RNA is created during transcription. Since this is already a map of an RNA transcript, transcription has already happened, and no transcription factors will bind at any location.

36. **D** At each level of the food chain, approximately 10% of energy is used to build biomass for the next level. The primary consumers would have approximately 298.7 g/m^2 of biomass, and the secondary consumers would have approximately 29.87 g/m^2 of biomass.

37. **A** The question states that the change affects the place where the substrate binds, which is defined as the active site. Position A is identified as the active site.

38. **D** Position B has nonpolar residues and is not the active site, since Position A is the active site. Therefore, (A) is not correct since the substrate does not bind at Position B. There is no information to indicate that (B) or (C) is incorrect. Nonpolar residues are known to be helpful for inserting into the lipid bilayer, which makes (D) a stronger answer.

39. **D** A homodimer is formed when more than one polypeptide subunit come together. This is defined as a quaternary structure, (D).

40. **C** The protein is known to be a homodimer, so it must have two polypeptides in it. Each polypeptide has one N-terminus, so this dimer protein would have two. They would not occur at the active site, so Position A should not be in the answer.

41. **B** If fats are insoluble in water, figuring out what type of mixture it forms when mixed with water would help target this property of lipids. The only option that discusses mixing with water is (B).

42. **A** Choice (A) is correct; glycogen is stored energy. Choice (B) describes energy use instead of storage, so eliminate it. Glycerol is not a carbohydrate, so eliminate (C). Hormones are not carbohydrates, so eliminate (D).

43. **B** Organic compounds are defined as molecules that contain carbon covalently bonded to hydrogen and are found in living organisms. Choice (B) is the only option that references carbon.

44. **A** Figure 1 is adenine, which is a nucleotide. Nucleotides are the monomers of nucleic acids, so the image represents (B) and (C). According to the information in the passage, nucleic acids relay genetic information, so (A) is the only answer that incorrectly describes the figure.

45. **A** A true-breeding white nettle would be homozygous dominant (WW). A true-breeding pink nettle would be homozygous recessive (ww). All offspring would inherit one dominant allele and one recessive allele, so they would all be heterozygous (Ww). Heterozygous nettles would display the dominant white phenotype.

46. **C** The F1 generation refers to the first generation of offspring. The previous question determined that all of the offspring from the first cross would be heterozygous. If the heterozygote offspring was taken and crossed with a true-breeding pink nettle, which would be homozygous recessive (ww), half of the offspring would be Ww and half would be ww. The ww offspring would have pink flowers.

47. **B** The heterozygote ($WwSs$) would be crossed with a nettle recessive for both traits ($wwss$). If the genes are not linked, the flowers should be equally white/ruffled, white/slotted, pink/ruffled, and pink/slotted. If the genes are linked, then some combinations of color/texture should be more common. Since a pattern is found in the offspring, and only some of those combinations are present, the traits are likely linked.

48. **B** The equations of Hardy-Weinberg ($p + q = 1$ and $p^2 + 2pq + q^2 = 1$) should be used. There are 5,300 plants, and 400 of them are pink plants (homozygous recessive). Therefore, the frequency of homozygous recessive genotypes = 400/5,300 = q^2 = 0.075 and thus q = 0.27.

49. **B** If you correctly solved the previous problem, you're halfway there, as you've already found that q = 0.27. However, here's that information again. The equations of Hardy-Weinberg ($p + q = 1$ and $p^2 + 2pq + q^2 = 1$) should be used. There are 5,300 plants, and 400 of them are pink plants (homozygous recessive). Therefore, the frequency of homozygous recessive alleles = 400/5,300 = q^2 = 0.075 and thus q = 0.27. If $p + q = 1$, the equation becomes $p + 0.27 = 1$, or p = 0.73. The frequency of heterozygotes is $2pq$, which becomes 2(0.73)(0.27) = 0.3942 = 39%.

50. **D** G-protein coupled receptors (GPCRs) are necessary to bring a signal from outside the cell to inside the cell. If the ligand is a small nonpolar molecule, then it can go through the membrane without trouble and does not need an extracellular receptor.

51. **B** According to the figure, the first step in the activation pathway was the binding of epinephrine, the external ligand. The scientist should aim to knock out this step first to confirm that this is the first step.

52. **C** The G-protein coupled receptor (GPCR) allows a message to be transmitted inside the cell without the ligand actually entering the cell. This is most like a drive-through window where the order is communicated to someone inside a restaurant, but the patron does not actually enter the building.

53. **B** The conformational change allows a message to be passed from the outside of the cell to the inside. The conformational change is a sign that the ligand is bound. The ligand does not bind to the membrane, so (A) is incorrect. A conformational change would not make an attachment more solid, which rules out (C). The ligand triggers binding of the GTP-protein, so (D) is incorrect.

AP Biology Premium Prep

54. **C** According to the table, the average carrot mass was stable around 60 grams, but the carrot length decreased by almost 4 cm over time.

55. **C** If a weevil lived at 17 cm, the carrots that are longer than 17 cm would not survive. The shorter carrots would be untouched by the weevils and have more evolutionary fitness. The shorter carrots would be naturally selected for over time, and the short trait would become more prevalent in the population.

56. **B** By removing the weevil population, any long carrots would have the chance to thrive, so eliminate (C) because it doesn't include Statement I. Additionally, by crossing the longest carrots, the genotypes that promote carrot length will be selected for, so eliminate (A), which doesn't include Statement II. Since all of the carrots are longer than 10 cm, a predator at 10 cm would still be able to prey on the carrots, which eliminates Statement III.

57. **B** If the carrot became shorter, the weevil population would still survive as long as the radish supply was sufficient. Without the radishes, the weevils that lived at those depths would die, and only the weevils that can live at depths where some food is found will survive, (B). Carrots grow to a depth of 17 cm, so the population of weevils might be selected to live at 17 cm. If only a small percentage of weevils died from a virus, this shouldn't change the preferred depth of the entire population.

58. **B** The table shows that the weevil predation is a significant selective pressure on carrot length. Length determines whether or not a carrot is able to prosper in its environment and display a high level of fitness.

59. **D** Sexual selection is defined as natural selection specifically for a mate of the opposite sex. The dance would increase mate attraction, but the predator would also decrease the health of the same mates. The sexual selection and the predation are selective pressures acting in opposing ways at the same time, and both are important.

60. **B** Bacteria reproduce through asexual reproduction, so they don't have "parents" as mentioned in (A). Other species are not mentioned as in (C), and the bacteria still fill the same niche before and after the resistance trait is developed as mentioned in (D).

Section II: Free Response

Short student-style responses have been provided for each of the questions. These samples indicate an answer that would get full credit, so if you're checking your own response, make sure that the actual answers to each part of the question are similar to your own. The structure surrounding them is less important, although we've modeled it as a way to help organize your own thoughts and to make sure that you actually respond to the entire question.

Note that the rubrics used for scoring periodically change based on the College Board's analysis of the previous year's test takers. We've done our best to approximate their structure, based on our institutional knowledge of how past exams have been scored and on the information released by the test makers.

Our advice is to over-prepare. Find a comfortable structure that works for you, and really make sure that you're providing all of the details required for each question. Also, continue to check the College Board's website, as they may release additional information as the test approaches. For further help, especially if you're worried that you're not being objective in scoring your own work, ask a teacher or classmate to help you out. Good luck!

Question 1

A. Bacterial communities containing different species of bacteria are no different than other communities with multiple species. **Explain** what factors contribute to a carrying capacity. (2 points)

Carrying capacity is determined by the available resources, including food/nutrients; a safe, healthy space to live in; water; sunlight; etc. Anything that is required but is in short supply contributes to the carrying capacity. When populations are competing for resources, there is a finite population that the environment can sustain, and that is the carrying capacity.

- 1 point for explaining that carrying capacity is **the maximum number of individuals of a species that a habitat can support**

- 1 point for explaining how limits on resources, such as food/nutrients, shelter, water, etc., **limit the number of individuals an environment can sustain**

B. Two important controls were used in this experiment. In addition, the experiment includes a 7-day waiting period between when the bacterial transplants take place.

   i. **Identify** the two controls used in this experiment. (2 points)

One important control is the use of a clean swab. Another control is to deposit the interdigital swab onto another interdigital space.

- 1 point for identifying a **clean swab** as a control

- 1 point for identifying **moving the interdigital swab to another space**, either an alternate skin site or the same type of skin site, as a control

   ii. **Justify** why the two controls and 7-day waiting period are necessary. (1 point)

These are important to confirm that the process of applying the swab to the new area is not a tainted or contaminated process. The 7-day waiting period is important to allow the new bacteria to compete a little with the existing bacteria. Then we can better judge which bacteria is going to survive in the new site.

- 1 point for justifying the 7-day waiting period as **providing time for the transplanted bacteria to compete with the already existing bacteria** AND the two controls as **confirming that the process is not contaminated**

C.

   i. **Interpret** the effect of the transplants between the interdigital web space and the buttock area. (1 point)

When deposited on the buttock area, the bacteria community already on the buttock did not get changed much. The existing bacteria still made up the community. This means the interdigital bacteria didn't survive well there or could not compete with the existing bacteria.

- 1 point for interpreting the effect of the transplant on the buttock area as having **little to no change**

AP Biology Premium Prep

    ii. **Interpret** the effect of the transplants between the interdigital web space and the external auditory canal area. (1 point)

When deposited on the external auditory canal area, it looked like the new bacteria made an impact and were competing with the existing bacteria. The external auditory canal displayed two new types of bacteria that weren't initially present.

- 1 point for interpreting the effect of the transplant on the external auditory canal area as **causing two new types of bacteria to start growing** that weren't initially present

    iii. **Interpret** the effect of the transplants between the interdigital web space and the plantar heel area. (1 point)

When deposited on the plantar heel area, it looked like the new bacteria made an impact and were competing with the existing bacteria. The plantar heel area seemed to be recolonized by the bacteria from the interdigital web space, as the relative composition of all four types of bacteria changed significantly.

- 1 point for interpreting the effect of the transplant on the plantar heel area as causing the **relative composition of all four types to change**

D. If each deposition site is pretreated with known antibiotics, **explain** how this would likely affect the results. (2 points)

Antibiotics destroy bacteria. If the new sites had been treated with antibiotics, this would have wiped out the existing bacteria. If the antibiotics were still present, then they would kill the newly deposited bacteria as well. Either way, the skin sites would be clear of the existing bacteria and ready to receive the newly deposited bacteria, like a clean slate. This would make the results show a population more like the interdigital than the existing bacteria at those sites.

- 1 point for explaining how antibiotics work in terms of **killing bacteria**
- 1 point for explaining how the results would change in response to antibiotics and cause there to be a **greater proportion of transplanted bacteria**

Question 2

A.

    i. **Describe** how CO_2 is a measure of respiration. (1 point)

CO_2 is a byproduct of cellular respiration, which is why we need to get rid of it when we exhale. Living things, like crickets, make more CO_2 when they are performing more cellular respiration (with the processes glycolysis, the Krebs cycle, and the electron transport chain). CO_2 is a byproduct of the Krebs cycle.

- 1 point for describing that CO_2 **is a product of cellular respiration** AND can therefore **be used to measure the rate of respiration**

    ii. **Explain** why temperature might affect the rate of respiration. (1 point)

Temperature might affect respiration because the crickets may change their behavior levels with the temperature. This makes them use more oxygen and energy and need more respiration when they are more active. Also, certain enzymes work better at different temperatures, so this might increase the rate of respiration or decrease the rate of respiration, depending on how the temperature impacts enzyme efficiency.

52 | For more free content, visit PrincetonReview.com

- 1 point for explaining that **temperature may cause increases or decreases in cricket behavior** OR that **temperature causes increases and decreases in enzyme activity**

B. On a set of axes, **construct** an appropriately labeled graph to demonstrate the CO_2 levels with changing temperature. Please include all temperatures, timepoints, and error bars. (4 points)

- 1 point for accurate graphing of the data
- 1 point for appropriate axes ranging from approximately **0 to 2500** on the y-axis AND approximately **0 to 12** on the x-axis
- 1 point for an appropriate title relating CO_2 **to Temperature**, appropriate labels of CO_2 on the y-axis AND **Time** on the x-axis, AND a **key** for the **three temperatures**
- 1 point for accurate graphing of error bars

C. **Identify** which temperature was associated with the highest respiration rate in the crickets. (1 point)

The highest respiration rate was found at 35°C, as that showed the greatest increase over time.

- 1 point for identifying the highest respiration rate as **35°C**

D.

　i. **Predict** what would happen to the CO_2 levels in the chamber with an additional 5°C increase above the highest temperature. (1 point)

With a 5°C increase in temperature, the CO_2 levels would increase to approximately 2100–2200 ppm.

- 1 point for predicting CO_2 levels of **2100–2200 ppm**

AP Biology Premium Prep

    ii.   **Justify** your prediction with evidence from the data. (1 point)

This is justified because the CO_2 levels increased by approximately 500 ppm each time the temperature increased by 15°C. This means that a 5°C increase in temp would increase the CO_2 levels by about 166 ppm.

- 1 point for justifying the prediction using SPECIFIC numbers from the graph AND/OR Table 1

Question 3

A. **Explain** how changing a single nucleotide can change the sequence of the protein. (1 point)

Changing a nucleotide will change the codon, which is the group of three nucleotides that gets read by the ribosome when the protein is made. If the codon changes, it might change the amino acid that gets added to the protein. It could also make the chain shorter if it changes to a stop codon.

- 1 point for explaining that **a single nucleotide can change a codon** OR **terminate a sequence early**

B. Given what is known about the neuraminidase enzyme, **describe** what is measured to provide the results in Figure 2 and Figure 3. (1 point)

The neuraminidase enzyme cleaves a tether that holds the viral particle to the membrane. To measure the "neuraminidase activity" for Figure 2 and Figure 3, the cleavage action of the enzyme must be measured.

- 1 point for describing that the **influenza neuraminidase (NA) protein cleaves a tether** AND that **the cleavage** is what will be measured

C. **Predict** whether an NA_H virus with an N234Y substitution will be a rapidly spreading or slowly spreading virus. (1 point)

It will be a slowly spreading virus.

- 1 point for predicting that the virus is **slow spreading**

D. **Justify** your prediction. (1 point)

The N234Y causes the enzyme activity to decrease, and this would mean that the viral particles do not get released. This means that the virus should not spread quickly. NA_H has high enzyme activity and spreads quickly, but the N234Y version should not spread quickly.

- 1 point for justifying the answer to 3c in relation to how **the N234Y substitution will decrease enzyme activity, causing fewer virus particles to be released**

Question 4

A. **Identify** which cellular process thylakoids are associated with. (1 point)

Thylakoids are associated with photosynthesis.

- 1 point for identifying the process as **photosynthesis**

B. **Explain** why ions in the hydroponic tank cannot pass into the roots by simple diffusion. (1 point)

The roots, like all cells, have a phospholipid bilayer membrane. This membrane is not permeable to polar or charged things. The inner hydrophobic space prevents things like ions from diffusing through. Ions must enter by facilitated diffusion or sometimes active transport using a pump.

- 1 point for explaining that all cells have a **phospholipid bilayer membrane** which will **not allow charged ions to pass through**

C. **Predict** the immediate effect this would have on the cell's ability to produce reactants necessary for carbon fixation. (1 point)

The reactants for carbon fixation in the Calvin cycle would not be produced.

- 1 point for predicting that the reactants would **NOT be produced**

D. **Justify** your prediction. (1 point)

The permeability of the thylakoid membrane would make the hydrogen ions able to flow freely from one side to the other. Without the ability to create a concentration gradient with hydrogen ions, the electron transport chain would fail. This would prevent production of ATP and NADPH, both of which are required for carbon fixation in the Calvin cycle.

- 1 point for justifying that the increased permeability will make it **impossible to create a concentration gradient**, which will **prevent ATP and NADPH from being produced**

Question 5

A. **Describe** what is represented by a node in a phylogenetic tree. (1 point)

A node represents a common ancestor. This means that the species that branch from that point had a common ancestor and were once the same species. That species got split, and each group evolved differently.

- 1 point for describing a node as a **common ancestor** that **branched into multiple species**

B. **Identify** two of the viruses that are LEAST similar to GGE. (1 point)

Two of the viruses with the least similarity to GGE are YF and SRE.

- 1 point for identifying TWO of the three viruses that are least similar: **YF, TYU,** AND/OR **SRE**

C. In 50 years, new data have identified that TBEwest(Hypr) is no longer in circulation, but it seems to have evolved in two directions: TBEwest1 and TBEwest2. Using the empty tree below, correctly **indicate** TBEwest(neu), TBEwest(Kuml

B. **Explain** why the lines of the graph never exceed 100. (1 point)

The graph is showing the percentages of moths of each color. 100% would represent the entire population of moths. It cannot be higher than 100%. The two colors of moths should always add up to 100% since they are the only possible colors of moth.

- 1 point for explaining that the *y*-axis is displaying **percentage of moth population** AND that the **two moth percentages should always add up to 100%**

C. **Predict** during which years there was a drought. (1 point)

There was likely a drought from approximately 1976–1988. The beginning of the drought seemed to happen quickly in 1976 when the green moths, who prosper in wet environments, dramatically decreased. The end of the drought is harder to predict, but the green moths appear to make a comeback starting around 1988, and the numbers approach pre-drought levels of approximately 50%, suggesting the drought conditions might be disappearing.

- 1 point for predicting that a drought occurred from **approximately 1976–1988**

D. **Explain** how natural selection led to the rise of the beige moth in the late 1970s. (1 point)

Natural selection caused the beige moths to become more plentiful because they survive better when the grass turns brown from the drought. This is probably because the birds cannot find them to eat them as well as the green moths, which are not camouflaged against the grass because the grass is dry and brown and not green. As the green moths get eaten, they do not reproduce, and the population becomes filled with a higher percentage of beige moths.

- 1 point for explaining how **natural selection favors the beige moths in times of drought** because they are **better able to camouflage from predators and therefore live to reproduce in the brown grass**

HOW TO SCORE PRACTICE TEST 1

Section I: Multiple Choice

Section II: Free Response

Composite Score

Composite Score Range	AP Score
57–68	5
45–56	4
36–44	3
21–35	2
0–20	1

Part III
About the AP Biology Exam

- The Structure of the AP Biology Exam
- The AP Biology Exam Is Hybrid Digital
- How the AP Biology Exam Is Scored
- Overview of Content Topics
- Breakdown of Free-Response Questions
- How AP Exams Are Used
- Other Resources
- Designing Your Study Plan

Calculators

Students may use the built-in Desmos graphing calculator in the Bluebook testing app during the exam. Students are also allowed to use a four-function (with square root), scientific, or graphing calculator. However, be aware that some models have unapproved features (like keyboards, styluses, or wireless capability) and are not permitted, so please check the College Board website. You can bring two calculators in case one is rejected (or not functioning).

Take Your Digital Tests Online

Register your book at princetonreview.com/prep to access your Digital Practice Tests via your Student Tools.

We've Got You Covered!

For more information on strategy and testing features for the digital AP exam, please see our "How to Approach Digital Testing for AP Students" PDF online via your Student Tools.

THE STRUCTURE OF THE AP BIOLOGY EXAM

The AP Biology Exam is three hours long and is divided into two sections: Section I (multiple-choice questions) and Section II (free-response questions).

Section I consists of 60 questions. You will have 90 minutes to complete this section.

Section II involves free-response questions. You'll be presented with two long-form free-response questions and four short-form free-response questions touching upon key issues in biology. You'll be given 90 minutes to answer all six questions.

If you're thinking that this sounds like a heap of work to try to finish in three hours, you're absolutely right. How can you possibly tackle so much science in so little time? Fortunately, there's absolutely no need to. As you'll soon see, we're going to ask you to leave a small chunk of the test blank. Which part? The parts you don't like. This selective approach to the test, which we call "pacing," is probably the most important part of our overall strategy. But before we talk strategy, let's look at the topics that are covered by the AP Biology Exam.

THE AP BIOLOGY EXAM IS HYBRID DIGITAL

While the multiple-choice section of AP Biology is administered digitally via the College Board's Bluebook testing app, the free-response questions are viewed digitally, but your answers must be written in the paper exam booklet given to you and returned for scoring. Bring a pencil or pen with black or dark blue ink to the exam. You may want to bring more than one in case your pencil tip breaks or your pen runs out of ink.

For the digital portion of the AP Biology Exam, access to the Bluebook app is available via Windows, Mac laptops/desktops, iPads, tablets, and Chromebooks. If your computer or tablet is owned or managed by your school, your school official will likely already have this installed. The test cannot be taken on a smartphone.

The Bluebook app allows you to annotate and highlight texts, eliminate answers, and flag questions for review later. You will also have access to scratch paper to plan responses to questions. Unlike other digital exams like the SAT, the digital AP exam will NOT be adaptive: this means that the difficulty of questions will not change depending on how well you do on earlier sections of the test.

It is important to note that the digital AP Biology Exam includes the same number of sections, types of questions, and timing as the previous paper exam. This means the strategies used for success on paper versions of the exam are just as applicable to the digital version. So don't worry, the practice tests in this book will still prepare you for test day and digital practice tests are available via your Student Tools.

HOW THE AP BIOLOGY EXAM IS SCORED

For your reference, here is the College Board's score distribution data from the 2024 AP Biology Exam administrations.

Score	2024 Percentage	Credit Recommendation	College Grade Equivalent
5	16.8%	Extremely Well Qualified	A
4	23.1%	Well Qualified	A–, B+, B
3	28.4%	Qualified	B–, C+, C
2	21.7%	Possibly Qualified	—
1	10%	No Recommendation	—

*Data from The College Board, May 2024 AP Exam administrations

Remember that colleges' rules may vary when it comes to granting credit for AP courses. You should contact the individual admissions departments to find out what score you need on the exam to ensure you'll be given credit.

For the AP Biology Exam, your scores on the multiple-choice section and free-response section are each worth 50 percent of your final score. On the multiple-choice section, your total score is based on the number of questions answered correctly, and you do not lose any points for incorrect answers. Unanswered questions do not receive points. The free-response section is graded on a separate point system. Your scores are tallied to determine your total free-response score. The free-response score is then combined with your multiple-choice score and weighted to figure out where your score falls within the standard AP scoring scale (1 to 5).

Go Online!
Check out us out on YouTube for test taking tips and techniques to help you ace your next exam at youtube.com/ThePrincetonReview.

Looking to Guarantee a 4 or 5?

We now offer one-on-one tutoring for a guaranteed 5 or an online course for a guaranteed 4 on the AP Biology Exam. For information on rates and availability, visit PrincetonReview.com/college/ap-test-prep.

OVERVIEW OF CONTENT TOPICS

The AP Biology Exam covers these four Big Ideas.

These are the *official* Big Ideas:

- **Big Idea 1: Evolution**—The process of evolution drives the diversity and unity of life.
- **Big Idea 2: Energetics**—Biological systems use energy and molecular building blocks to grow, to reproduce, and to maintain dynamic homeostasis.
- **Big Idea 3: Information Storage and Transmission**—Living systems store, retrieve, transmit, and respond to information essential to life processes.
- **Big Idea 4: Systems Interactions**—Biological systems interact, and these systems and their interactions exhibit complex properties.

As you read this book, think about how these themes fit with various areas of biology:

- Building blocks/hierarchy
- Responses to stimuli
- Organization
- Structure/function
- Communication
- Relationships
- Disruptions and consequences
- Critical thinking

To fully understand the four big ideas, a solid grasp of eight content units is required. The following is a list of units/topics and the weight of each unit on the exam.

- Chemistry of Life (8–11%)
    - Important properties of water
    - pH
    - Carbohydrates
    - Proteins
    - Lipids
    - Nucleic acids
    - Origins of life

- Cell Structure and Function (10–13%)
    - Prokaryotic and eukaryotic cells
    - Organelles
    - Membranes and transport
    - Cell junctions
    - Cell communication

- Cellular Energetics (12–16%)
  - Change in free energy
  - Enzymes
  - Coupled reactions and ATP
  - Photosynthesis
  - Cellular respiration (glycolysis, Krebs, oxidative phosphorylation)
  - Fermentation

- Cell Communication and Cell Cycle (10–15%)
  - Cell cycle
  - Mitosis
  - Meiosis

- Heredity (8–11%)
  - Mendelian genetics
  - Inheritance patterns

- Gene Expression and Regulation (12–16%)
  - DNA and genome structure
  - Transcription
  - Translation
  - Gene regulation
  - Mutation
  - Biotechnology

- Natural Selection (13–20%)
  - Natural selection
  - Evidence of evolution
  - Phylogenetic trees
  - Impact of genetic variation
  - Speciation
  - Hardy-Weinberg equilibrium

- Ecology (10–15%)
  - Behavior and communication
  - Food webs and energy pyramids
  - Succession
  - Communities and ecosystems
  - Global issues

This might seem like an awful lot of information, but for each topic, there are just a few key facts you'll need to know. Your biology textbooks may go into far greater detail about some of these topics than we do. That's because they're trying to teach you "correct science," whereas we're aiming to improve your scores. Our science is perfectly sound; it's just cut down to size. We've focused on crucial details and given you only what's important. Moreover, as you'll soon see, our treatment of these topics is far easier to handle.

The AP Biology Exam not only tests your content knowledge, but it also tests how you apply that knowledge during scientific inquiry. Simply put, the test's authors are testing whether you can design and/or think critically about experiments and the hypotheses, evidence, math, data, conclusions, and theories therein. There are six broad science practices that are tested. The weight that each practice holds within the multiple-choice section is given.

> - **Science Practice 1: Concept Explanation:** Explain biological concepts and processes presented in written format. (25–33%)
> - **Science Practice 2: Visual Representations:** Analyze visual representations of biological concepts and processes. (16–24%)
> - **Science Practice 3: Questions and Methods:** Determine scientific questions and methods. (8–14%)
> - **Scientific Practice 4: Representing and Describing Data:** Represent and describe data. (8–14%)
> - **Scientific Practice 5: Statistical Tests and Data Analysis:** Perform statistical tests and mathematical calculations to analyze and interpret data. (8–14%)
> - **Science Practice 6: Argumentation:** Develop and justify scientific arguments using evidence. (20–26%)

BREAKDOWN OF FREE-RESPONSE QUESTIONS

Each of the six free-response questions will follow a slightly different outline.

Questions 1 and 2 are long questions worth 8–10 points each.

Question 1: Interpreting and Evaluating Experimental Results

Students will be provided with a scenario and an accompanying graph and/or table.

- Part A (1–2 points): **Describe** and **explain** biological concepts, processes, or models.
- Part B (3–4 points): **Identify** experimental design procedures.
- Part C (1–3 points): Analyze data.
- Part D (2–4 points): Make and **justify** predictions.

Question 2: Interpreting and Evaluating Experimental Results with Graphing

This question is similar to question 1, except the student will be required to construct a data representation.

- Part A (1–2 points): **Describe** and **explain** biological concepts, processes, or models.
- Part B (4 points): **Construct** a graph, plot, or chart with confidence intervals or error bars.
- Part C (1–3 points): Analyze data.
- Part D (1–3 points): Make and **justify** predictions.

Questions 3–6 are short questions that are each worth 4 points.

Question 3: Scientific Investigation

Students will be provided with a lab investigation scenario.

- Part A (1 point): **Describe** biological concepts or processes.
- Part B (1 point): **Identify** experimental procedures.
- Part C (1 point): **Predict** results.
- Part D (1 point): **Justify** predictions.

Question 4: Conceptual Analysis

Students are provided a scenario describing a biological phenomenon with a disruption.

- Part A (1 point): **Describe** biological concepts or processes.
- Part B (1 point): **Explain** biological concepts or processes.
- Part C (1 point): **Predict** the causes or effects of a change in a biological system.
- Part D (1 point): **Justify** predictions.

Question 5: Analyze Model or Visual Representation

Students are provided a scenario accompanied by a visual model or representation.

- Part A (1 point): **Describe** characteristics of a biological concept, process, or model represented visually.
- Part B (1 point): **Explain** relationships between different characteristics of a biological concept or process represented visually.
- Part C (1 point): **Represent** results within a biological model.
- Part D (1 point) **Explain** how a biological concept or process represented visually relates to a larger biological principle, concept, process, or theory.

Question 6: Analyze Data

Students are provided data in a graph, table, or other visual representation.

- Part A (1 point): Describe data.
- Part B (1 point): Describe data.
- Part C (1 point): Use data to **evaluate** a hypothesis or prediction.
- Part D (1 point): **Explain** how experimental results relate to biological principles, concepts, processes, or theories.

HOW AP EXAMS ARE USED

Different colleges use AP Exams in different ways, so it is important that you go to a particular college's website to determine how it uses AP Exams. The three items below represent the main ways in which AP Exam scores can be used:

- **College Credit.** Some colleges will give you college credit if you score well on an AP Exam. These credits count toward your graduation requirements, meaning that you can take fewer courses while in college. Given the cost of college, this could be quite a benefit, indeed.
- **Satisfy Requirements.** Some colleges will allow you to "place out" of certain requirements if you do well on an AP Exam, even if they do not give you actual college credits. For example, you might not need to take an introductory-level course, or perhaps you might not need to take a class in a certain discipline at all.
- **Admissions Plus.** Even if your AP Exam will not result in college credit or allow you to place out of certain courses, most colleges will respect your decision to push yourself by taking an AP Course or an AP Exam outside of a course. A high score on an AP Exam shows comprehension of more difficult content than is taught in many high school courses, and colleges may take that into account during the admissions process.

Got a Question?
For answers to test-prep questions for all your tests and additional test taking tips, subscribe to our YouTube channel at youtube.com/ThePrincetonReview.

More Great Books
Check out The Princeton Review's college guidebooks, including *The Best 391 Colleges, Paying for College, The Complete Guide to College Essays*, and many more!

OTHER RESOURCES

There are many resources available to help you improve your score on the AP Biology Exam, not the least of which are your **teachers**. If you are taking an AP class, you may be able to get extra attention from your teacher, such as obtaining feedback on your essays. If you are not in an AP course, reach out to a teacher who teaches AP Biology and ask if the teacher will review your essays or otherwise help you with content.

Another wonderful resource is AP Students, the official student site of the AP Exams. The scope of the information at this site is quite broad and includes:

- the course description, which provides details on what content is covered and sample questions
- the AP Biology Exam free-response and scoring guidelines
- access to AP Classroom if you are enrolled in a course (teacher assistance required)
- free-response prompts from previous years and exam tips

The AP Students web page is apstudents.collegeboard.org.

Finally, The Princeton Review offers tutoring and small group instruction. Our expert instructors can help you refine your strategic approach and add to your content knowledge. For more information, call 1-800-2REVIEW.

If there are any late-breaking changes from the College Board after the printing of this book, we'll let you know about them in your free online Student Tools.

Go online to access two more AP Bio practice exams. Head over to PrincetonReview.com and register this book for a host of test prep resources, including extra AP Bio practice tests and SAT and ACT practice!

If you are a visual learner, check out our other books, *Fast Track Biology: Essential Review for AP, Honors, and Other Advanced Study* and *ASAP Biology: A Quick-Review Study Guide for the AP Exam*. These handy guides focus on the most tested content and present it in a friendly, illustrated fashion.

DESIGNING YOUR STUDY PLAN

In Part I, you identified some areas of potential improvement. Let's now delve further into your performance on Practice Test 1 with the goal of developing a study plan appropriate to your needs and time commitment.

Read the answers and explanations associated with the multiple-choice questions (starting at page 45). After you have done so, respond to the following questions:

- Review the bulleted list of topics on pages 62–63. Next to each topic, indicate your rank of the topic as follows: 1 means "I need a lot of work on this," 2 means "I need to beef up my knowledge," and 3 means "I know this topic well."

- How many days/weeks/months away is your exam?

- What time of day is your best, most focused study time?

- How much time per day/week/month will you devote to preparing for your exam?

- When will you do this preparation? (Be as specific as possible: Mondays and Wednesdays from 3:00 to 4:00 P.M., for example.)

- Based on the answers above, will you focus on strategy (Part IV), content (Part V), or both?

- What are your overall goals in using this book?

Use those answers to create a study plan. Start with the topics that need the most work and map out when you will study and what you will study.

Remember, your schedule may evolve along the way. If a certain time/location is not working for you, then try mixing it up. If you are struggling with a topic, perhaps try tackling it with a teacher, a tutor, or a classmate.

Part IV
Test-Taking Strategies for the AP Biology Exam

- Preview
1 How to Approach Multiple-Choice Questions
2 How to Approach Free-Response Questions
3 Using Time Effectively to Maximize Points

PREVIEW

Review your responses to the three questions on page 2 and then respond to the following questions:

- How many multiple-choice questions did you miss even though you knew the answer?

- On how many multiple-choice questions did you make a random guess?

- How many multiple-choice questions did you miss after eliminating some answers and guessing based on the remaining answers?

- Did you find any of the free-response questions easier/harder than the others—and, if so, why?

HOW TO USE THE CHAPTERS IN THIS PART

For the following Strategy chapters, think about what you are doing now before you read the chapters. As you read and engage in the directed practice, be sure to appreciate the ways you can change your approach. At the end of Part IV, you will have the opportunity to reflect on how you will change your approach.

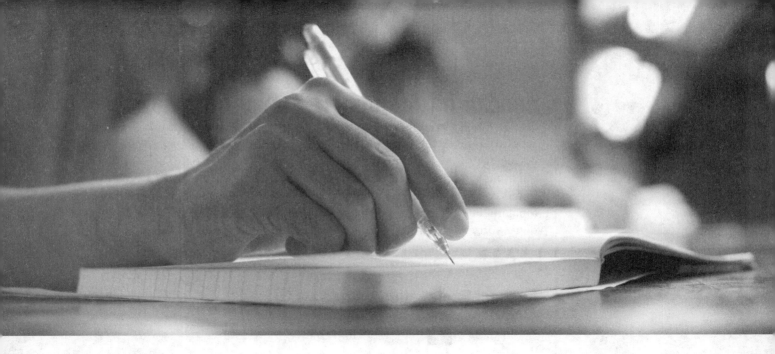

Chapter 1
How to Approach Multiple-Choice Questions

AP Biology Premium Prep

CRACKING THE MULTIPLE-CHOICE QUESTIONS

Section I consists of 60 multiple-choice questions that deal with an experiment, set of data, or with general knowledge of biology. For a breakdown of the average percentage of each unit/topic, please refer back to pages 62–63. Each of these questions will also test a specific Science Practice topic, the list of which can be found on page 64. On the following pages, we'll provide an example of how that topic might be tested in multiple-choice form. These are just model questions, so don't worry about the answers—we just want you to see how these might look in context, and with figures, diagrams, and/or charts.

Concept Explanation

1   🔖 Mark for Review

Electrophoresis is used to separate DNA fragments by their size and charge. DNA fragments are pipetted into an agarose gel well and covered with a buffer. An electrical current is applied to the gel and buffered solution. Which of the following best explains why DNA fragments migrate towards the positive pole through the agarose gel?

(A) The nitrogenous bases found in DNA have a positive charge.

(B) The phosphate groups found in DNA have a negative charge.

(C) The histone proteins which are intertwined with the DNA have a negative charge.

(D) The polar bonds on the deoxyribose molecules found in DNA have a slightly negative charge.

Concept Explanation questions require you to explain biological concepts, processes, and written models. These are your standard content knowledge questions, where you may be asked to apply prior knowledge to a specific context or situation. These questions typically provide a specific example, then ask you to identify or explain something more general about the example.

The most obvious feature of these questions is that they start with a few sentences of information, usually containing a specific example or phenomenon. Many of the questions start with "which of the following best describes/explains…" and ask you to identify something's function or explain the general science behind a situation.

72 | For more free content, visit PrincetonReview.com

Visual Representation

2 Mark for Review

Figure 1. Family pedigree of Marfan syndrome

Marfan syndrome is a genetic disorder that affects tissues in the body. People with Marfan syndrome are typically tall and thin and have scoliosis and high levels of flexibility in their joints. The pattern of incidence of Marfan syndrome in a specific family is represented in Figure 1.

Which of the following best describes the inheritance pattern for the disorder represented in Figure 1?

(A) X-linked recessive

(B) Autosomal recessive

(C) Autosomal dominant

(D) Incomplete dominant

Visual Representations questions require you to analyze pictorial depictions of biological concepts and processes. These questions are similar to Concept Explanation questions in that you may be asked to apply prior knowledge to a specific context or situation, except they will include a model, diagram, flow chart, or other representation instead of, or in addition to, a written description. These questions may also ask you to explain relationships between different topics or between specific concepts and larger biological principles and theories.

The most obvious feature of these questions is that they come with a graphic. Many of the questions start with "which of the following best describes/explains…" and reference a figure. You may be asked to identify the trend represented in a figure or for an explanation behind something in the figure.

Argumentation

| 3 | ☐ Mark for Review |

A mosquito species that is a carrier for dengue fever lives in a rainforest in which two species of anteaters cohabit. Species A is immune to dengue fever but species B is not. The mosquito species is the primary food source for a specific species of bat that also lives in the rainforest. Which of the following statements best predicts the observable consequence if roost destruction suddenly eliminates all of the bats?

(A) Anteater species A will experience a higher death rate.

(B) Anteater species B will experience a higher death rate.

(C) Anteater species B will develop resistant strains of dengue fever.

(D) Anteater species A will develop sensitive strains of dengue fever.

Argumentation questions require you to make and support claims using evidence, biological principles and theories, and/or evidence. You may be asked to provide reasoning to justify a claim, explain the relationship between experimental results and larger concepts, and predict the effects of a disruption in a biological system or model. Evidence could be provided as a general biological concept or process, a visual representation, and/or data. These questions typically provide a specific example, then ask you to make a claim or predict something about the example.

The most obvious feature of Argumentation questions is that they reference a source of data or a piece of information. While Concept Explanation questions also reference information, they are more focused on memorized biological facts instead of justification and evidence. Many of the questions start with "which of the following best predicts the consequence..." or "which of the following best supports the claim that...."

Questions 4 through 6 refer to the following.

Table 1. Effect of temperature on glucose production

Temperature	0°C	20°C	40°C	60°C	80°C
Rate of glucose production (mg/dL)/min	0	68.6	105	6.6	0
Color of solution	yellow	light green	dark green	very light green	yellow

Figure 1. Effect of heat on the structure of lactase

Normal protein　　　　　Denatured protein

Lactase is an enzyme found in many mammals that breaks down lactose, a sugar found in milk, into glucose through a hydrolysis reaction. Scientists hypothesized that lactase's efficiency is affected by temperature. To test this hypothesis, the scientists dissolved lactase tablets in 200 mL of water and mixed 1 mL of the resulting enzyme solution with 2 mL of milk at different temperatures. After 20 minutes, the scientists tested for the presence of glucose using test strips. The glucose test strips are yellow in the absence of glucose and turn increasingly darker shades of green when glucose is present. The results are shown in Figure 1.

Questions & Methods

4　Mark for Review

Which of the following would have provided the best control group for the investigation?

Ⓐ A sample of milk containing no enzyme solution

Ⓑ A sample of milk at a pH of 8

Ⓒ A sample containing orange juice instead of milk

Ⓓ A sample of milk at 100°C

Questions & Methods questions ask you to determine the methods best suited for an experiment. You may be asked to develop a testable question to prove or disprove an idea and/or develop hypotheses for an experiment. You may also be asked to identify experimental design components for experiments, such as independent and dependent variables and controls. You may be asked to propose the next steps for an experiment based on the methods used or the data collected.

The most obvious feature of Questions & Methods questions is the use of experimental design vocabulary, such as question, independent variable, dependent variable, and control. Many of the questions start with "which of the following questions will best guide the scientists towards…" or "which of the following best describes the independent variable/dependent variable/control of the experiment…."

Representing & Describing Data

5 — Mark for Review

Which of the following claims is best supported by the data?

(A) Lactase breaks down lactose most efficiently at low temperatures.

(B) Lactase breaks down lactose most efficiently at high temperatures.

(C) Lactase breaks down lactose most efficiently at moderate temperatures.

(D) Lactase efficiency is not affected by temperature.

Representing & Describing Data questions ask you to describe data from a table or graph, including identifying data points, describing trends and patterns, and describing relationships between variables. You may be asked to analyze a variety of representations, such as line graphs, bar graphs, logarithmic data, dual axes, histograms, box-and-whisker plots, and pie charts.

The most obvious feature of Representing & Describing Data questions is that they include an accompanying graph or data table. Many of the questions start with "which of the following claims is best supported by…" or "which of the following best describes the relationship between…."

Statistical Tests & Data Analysis

6 ☐ Mark for Review

The average rate of glucose production from 20°C to 60°C is closest to which of the following?

(A) 0 (mg/dL)/min

(B) 20 (mg/dL)/min

(C) 60 (mg/dL)/min

(D) 100 (mg/dL)/min

Statistical Tests & Data Analysis questions ask you to perform calculations, such as finding averages, rates, ratios, and percentages, or to draw conclusions from data. You may be asked to determine whether information is statistically significant or different based on confidence intervals or error bars, and you may be asked to perform chi-square statistics. While Argumentation questions also ask you to support claims with evidence, Statistical Tests & Data Analysis questions will focus more on the data than the scientific principles and theories behind the data.

The most obvious feature of Statistical Tests & Data Analysis questions is the associated numerical data and numbers in the answer choices. Many of the questions start with "The average/ratio/percent ... is closest to which of the following...."

> In case you were curious, the answers to the questions modeled on these pages are: 1 (B), 2 (C), 3 (B), 4 (A), 5 (C), 6 (C).

PACE YOURSELF

When you take a test in school, how many questions do you answer? Naturally, you try to answer all of them. You do this for two reasons: (1) your teacher told you to, and (2) if you left a question blank, your teacher would mark it wrong. However, that's not the case when it comes to the AP Biology Exam. Before we explain why, let's talk about timing.

One of the main reasons that taking the AP Biology Exam can be so stressful is the time constraint—90 seconds per multiple-choice question, 25 minutes per long essay, and 10 minutes per short essay. If you had all day, you would probably ace the test. We can't give you all day, but we can do the next best thing: we can give you more time for each question. How? By having you slow down and answer fewer questions.

Slowing down and doing well on the questions you do answer is the best way to improve your score on the AP Biology Exam. Rushing through questions in order to finish, on the other hand, will always hurt your score. When you rush, you're far more likely to make careless errors, misread, and fall into traps. Keep in mind that blank answers are not counted against you.

Let's Get Digital

Remember that the multiple-choice section of the AP Biology Exam will be administered in a digital format, so you will be taking them on the Bluebook app. It's not taken online, but it is taken digitally (Section I, that is). The test will be administered at your school on a desktop or laptop computer that has the Bluebook app installed. To prepare for this, it's best if you practice as much as you can in a digital environment so that you will be used to the look and feel of the test and know how to use the in-app Desmos calculator (if you wish to use that instead of your own calculator). Head to your online Student Tools (found at princetonreview.com/prep) and register your book to access the digital practice tests.

THE TWO-PASS SYSTEM

The AP Biology Exam covers a broad range of topics. There's no way, even with our extensive review, that you will know everything about every topic in biology. So what should you do?

Adopt what we call the two-pass system. The two-pass system entails going through the test and answering the easy questions first. Save the more time-consuming questions for later. (Don't worry—you'll have time to do them later!) First, read the question and decide if it a "now" or "later" question. If you decide this is a "now" question, answer it in the test booklet. If it is a "later" question, use the Mark for Review tool to add a little flag next to the question. Then keep on moving.

Once you've finished all the "now" questions, go back to the questions that you Marked for Review and start in on those ones. Start with the easier questions first. These are the ones that require calculations or that require you to eliminate the answer choices (in essence, the correct answer does not jump out at you immediately).

PROCESS OF ELIMINATION (POE)

On most tests, you need to know your material backward and forward to get the right answer. In other words, if you don't know the answer beforehand, you probably won't answer the question correctly. This is particularly true of fill-in-the-blank and essay questions. We're taught to think that the only way to get a question right is by knowing the answer. However, that's not the case on Section I of the AP Biology Exam. You can get a perfect score on this portion of the test without knowing a single right answer, provided you know all the wrong answers!

What are we talking about? This is perhaps the single most important technique in terms of the multiple-choice section of the exam. Let's take a look at the example below.

| 7 | ☐ Mark for Review |

Trees and shrubs have hard, woody stems and grow buds that persist above ground throughout the colder months. Which of the following best explains how gas is exchanged in a woody stem?

(A) Oxygen is taken in by lungs and carbon dioxide is released into the air.

(B) Oxygen, carbon dioxide, and water vapor pass through small pores called lenticels.

(C) Hormones and neurotransmitters are transferred between adjacent ganglia.

(D) Lentil beans provide small openings for gases to pass into the woody stem.

Now, if this were a fill-in-the-blank-style question, you might be in a heap of trouble. But let's take a look at what we've got. You see "woody stems" and "buds" in the question, which leads you to conclude that the question is about plants. Right away, you know the answer is not (A) or (C) because plants don't have lungs or ganglia. Now you've narrowed it down to (B) and (D). Notice that (B) and (D) look very similar. Obviously, one of them is a trap. At this point, if you don't know what "lentil beans" are, you have to guess. However, even if you don't know precisely what they are, it's safe to say that you probably know that lentil beans have nothing to do with gas exchange in plants. Therefore, the correct answer is (B).

> **Process of Elimination (POE)** is the best way to approach the multiple-choice questions.

Even if you don't know the answer right off the bat, you'll surely know that two or three of the answer choices are not correct. What then?

Active Guessing

The College Board, the service that develops and administers the exam, tells you that random guessing will not affect your score. This is true. There is no guessing penalty on the AP Biology Exam. For each correct answer you'll receive one point, and you will not lose any points for each incorrect answer.

Although you won't lose any points for wrong answers, you should guess strategically by getting rid of the incorrect answer choices. The moment you've eliminated a couple of answer choices, your odds of getting the question right, even if you guess, are far greater. If you can eliminate as many as two answer choices, your odds improve enough that it's in your best interest to guess.

WORD ASSOCIATIONS

Another way to rack up the points on the AP Biology Exam is by using word associations in tandem with your POE skills. Make sure that you memorize the words in the Key Terms lists throughout this book. Know them backward and forward. As you learn them, make sure you group them by association, since the College Board is bound to ask about them on the AP Biology Exam. What do we mean by "word associations"? Let's take the example of mitosis and meiosis.

You'll soon see from our review that there are several terms associated with mitosis and meiosis. Synapsis, crossing-over, and tetrads, for example, are words associated with meiosis but not mitosis. We'll explain what these words mean later in this book. For now, just take a look.

---

8   ▢ Mark for Review

In mitosis, plant cytokinesis differs from animal cytokinesis. Scientists studying the process have hypothesized that the Golgi apparatus sends cellulose in vesicles to form a cell plate between plant daughter cells. Which of the following questions will help guide the scientists toward a direct test of their hypothesis?

(A) In what types of cells do homologous chromosomes exchange DNA through crossing-over?

(B) Are tetrads formed in the center of the cell as a result of plant cytokinesis?

(C) When homologous chromosomes pair up through the process of synapsis, does a cell plate form?

(D) Does cytokinesis still occur in plant cells when Golgi vesicles are inhibited from traveling to the center of the cell?

This might seem like a difficult problem. But let's think about the associations we just discussed. The question asks us about mitosis. However, (A), (B), and (C) all mention events that we've associated with meiosis. Therefore, they are out. Without even racking your brain, you've managed to find the correct answer: (D). Not bad!

Once again, don't worry about the science for now. We'll review it later. What is important to recognize is that by combining the associations we'll offer throughout this book and your strategic POE techniques, you'll be able to rack up points on problems that might have seemed difficult at first.

> **Bonus Tips and Tricks…**
> Check us out on YouTube for additional test-taking tips and must-know strategies at youtube.com/ThePrincetonReview.

MNEMONICS—OR THE BIOLOGY NAME GAME

One of the big keys to simplifying biology is the organization of terms into a handful of easily remembered packages. The best way to accomplish this is by using mnemonics. Biology is all about names: the names of chemical structures, processes, theories, and so on. How are you going to keep them all straight? A mnemonic, as you may already know, is a convenient device for remembering something.

For example, one important issue in biology is taxonomy—that is, the classification of life-forms, or organisms. Organisms are classified in a descending system of similarity, leading from domains (the broadest level) to species (the most specific level). The complete order runs as follows: domain, kingdom, phylum, class, order, family, genus, and species. Don't freak out yet. Use a mnemonic to help you:

King Philip of Germany decided to walk to America. What do you think happened?

Dumb	→	**D**omain
King	→	**K**ingdom
Philip	→	**P**hylum
Came	→	**C**lass
Over	→	**O**rder
From	→	**F**amily
Germany	→	**G**enus
Soaked	→	**S**pecies

Learn the mnemonic and you'll never forget the science!

Mnemonics can be as goofy as you like, as long as they help you remember. Throughout this book, we'll give you mnemonics for many of the complicated terms we'll be seeing. Use ours, if you like them, or feel free to invent your own. Be creative! Remember: the important thing is that you remember the information, not how you remember it.

Practice Makes Perfect
This question type may not appear on the redesigned AP Biology Exam. However, because the POE skills they teach remain valuable for solving all other problems, our sample tests will, for the time being, continue to feature these questions for your benefit.

Outsmart the Tricky Questions
EXCEPT and NOT questions are considered more difficult than other question types. Here's where POE comes in handy!

Remember Your Resources!
The AP Biology Equations and Formulas sheet is on pages 333–334 and in your online Student Tools. You'll have a copy included with your test, so get in the habit of referring to it while you practice.

IDENTIFYING "EXCEPT" QUESTIONS

You might encounter some EXCEPT/NOT/LEAST questions. With this type of question, you must remember that you're looking for the wrong (or the least correct) answer. The best way to go about these is by using POE.

More often than not, the correct answer is a true statement, but is wrong in the context of the question. However, the other three tend to be pretty straightforward. Cross off the three that apply, and you're left with the one that does not. Here's a sample question.

All of the following are true statements about gametes EXCEPT

(A) they are haploid cells

(B) they are produced only in the reproductive structures

(C) they bring about genetic variation among offspring

(D) they develop from polar bodies

If you don't remember anything about gametes and gametogenesis, or the production of gametes, this might be a particularly difficult problem. We'll see these again later on, but for now, remember that gametes are the sex cells of sexually reproducing organisms. As such, we know that they are haploid and are produced in the sexual organs. We also know that they come together to create offspring.

From this very basic review, we know immediately that (A) and (B) are not our answers. Both of these are accurate statements, so we eliminate them. That leaves us with (C) and (D). If you have no idea what (D) means, focus on (C). In sexual reproduction, each parent contributes one gamete, or half the genetic complement of the offspring. This definitely helps vary the genetic makeup of the offspring. Choice (C) is a true statement, so it can be eliminated. The correct answer is (D).

Don't sweat it if you don't recall the biology. We'll be reviewing it in detail soon enough. For now, remember that the best way to answer these types of questions is to spot all the right statements and cross them off. You'll wind up with the wrong statement, which happens to be the correct answer.

Chapter 2
How to Approach Free-Response Questions

CRACKING THE FREE-RESPONSE QUESTIONS

Section II of the exam is made up of six free-response questions. Although each of these questions generally has four parts, labeled A, B, C, and D, each of which is to be answered separately, the first two questions tend to be longer, and may have the bolded task words embedded in between text and figures. The four shorter questions tend to have all the bolded task words at the end. You're largely expected to answer in complete sentences. In a few instances, you may be asked instead to show your mathematical work, or to create a diagram, graph, or table that models something from the question.

Ultimately, the best way to rack up points in this section is to give the test readers exactly what they're looking for. However, unlike the multiple-choice, where there is only one valid response—a letter that either matches or doesn't—these free-response readers will accept any answer that accurately fits the checklist that they have been given. (The "accurate" part is important. If you just list a bunch of incorrect things along with the correct response, the reader may assume you're just guessing.) In short, be short! Don't waste time including material that hasn't been asked for. Support your statements as needed, and write in full sentences (as opposed to outlines or bullet points). Use the following strategies to focus on providing specific, complete answers.

KNOW WHAT YOU NEED

The first thing you should do is take less than a minute to skim all of the questions and put them into your own personal order of difficulty from easiest to toughest. Once you've decided the order in which you will answer the questions (easiest first, hardest last), you can begin to formulate your responses. Your first step should be a more detailed assessment of each question.

The most important advice we can give you is to read each question at least twice. As you read the question, focus on key words, especially the bolded direction words. Almost every essay question contains a direction word. As you read through those outlines of what each question contains, you will recognize various direction words. Some examples of direction words are **calculate**, **explain**, **identify**, and **predict**. These words tell you exactly what the question is looking for, and they're exactly what you should do. For instance, a question that asks you to support a claim expects you to actually provide reasoning and/or evidence that backs it up. By contrast, a question that asks you to construct, draw, or represent data may just want you to include a graph or build a table, no further explanation necessary.

Many students lose points because they either misread the question or fail to do what's asked of them.

Remember the Questions

For a full list of the six types of questions and how they're scored, refer to pages 64–66.

For a complete list of the direction words (aka task verbs) used in the exam and their definitions, visit apstudents. collegeboard.org/courses/ ap-biology/exam-tips.

FIND WHAT YOU NEED

Now that you know what you need, you have to figure out where your answers are coming from. There are only two options, so this isn't as tricky as it might sound. Some parts can be answered based entirely on what you've learned in the course thus far; others, however, will require you to cite evidence or use information from the passage and/or graph that precedes the question.

Spend about two minutes per question re-familiarizing yourself with any key terms, definitions, or processes that you'll need to answer the question.

USE WHAT YOU NEED

The multiple parts of each question may connect to each other, but they're answered individually. Label the information that you've just set aside so that you know you've got enough to get all the available points for each part. If you've brainstormed more examples than the question is asking you for, choose only those that you feel strongest talking about.

Spend about a minute here organizing your thoughts, making sure that you're addressing as many of the bolded verbs as you can. You want to answer every part of a question so as to maximize your score, but if there's something you just don't know, first focus on the other parts. Then, time permitting, go back and look at any unanswered parts: a well-written, educated guess may get you partial credit, even if you're not sure about it.

ANSWER EACH PART OF THE QUESTION SEPARATELY

Unless one part specifically references an earlier one, treat each part as its own question. This shouldn't be difficult to remember. Each part should already have its own space in which to write or draw a response, and if not, you can feel free to break between parts by inserting headers—part A, part B, etc.

The key here is to make sure you answer as many parts as you can and to not get bogged down in any one part, because each part has a maximum value. For the first two long questions, you can score at most 5 points for a part (and if this is the case, the others will only be worth 1 to 2 points). For the four short questions, each is worth a single point. Ask yourself: is the extra point you might be able to squeak out of a long response worth the individual points you'll miss out on elsewhere if you run out of time on the short responses?

Room to Write
On exam day, you will view the free-response questions in the Bluebook testing app and handwrite your responses in a paper exam booklet. You can download a sample AP Biology booklet at apcentral.collegeboard. org/exam-administration- ordering-scores/digital-ap- exams/hybrid-digital. You can use scrap paper for the Free Response Questions in this book. For the practice tests, we also provide printable free-response booklets, which are available in your Student Tools via PrincetonReview.com/ prep.

You don't need a perfectly written response to get the maximum number of points, either, so don't waste your time with fancy introductions like "It was the best of experiments, it was the worst of experiments." Just leap right into your response. Don't even sweat grammar or spelling errors; these can hurt you only if they're so bad that they seriously impair the reader's ability to understand you.

Know How to Label Diagrams and Figures

Some responses of the **represent** variety may require you to answer with a figure as opposed to text. (The same is true for any math-based **calculate** parts.) The key here is to clearly and correctly show your work and understanding, so let's briefly discuss the important elements in setting up a graph. The favorite type of graph on the AP Biology Exam is the **coordinate graph**. The coordinate graph has a horizontal axis (*x*-axis) and a vertical axis (*y*-axis).

The *x*-axis usually contains the **independent variable**—the thing that's being manipulated or changed. The *y*-axis contains the **dependent variable**—the thing that's affected when the independent variable is changed.

Now let's look at what happens when you put some points on the graph. Every point on the graph represents both an independent variable and a dependent variable.

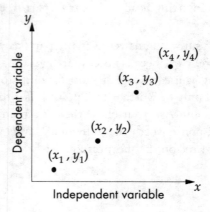

Once you draw both axes and label the axes as *x* and *y*, you can plot the points on the graph. We're just going to look at part B, which asks us to design an experiment.

> Enzymes are biological catalysts.
>
> B. **Construct** a graph that demonstrates the influence of temperature, substrate concentration, or pH on the activity of an enzyme.

Let's set up a graph that shows the results of an experiment examining the relationship between pH and enzyme activity. Notice that we've chosen only one factor here, pH. We could have chosen any of the three. Why did we choose pH and not temperature or substrate concentration? Well, perhaps it's the one we know the most about.

What is the independent variable? It is pH. In other words, pH is being manipulated in the experiment. We'll therefore label the *x*-axis with pH values from 0 through 14.

What is the dependent variable? It's the enzyme activity—the thing that's affected by pH. Let's label the *y*-axis "Rate of Reaction." Now we're ready to plot the values on the graph. Based on our knowledge of enzymes, we know that for most enzymes the functional range of pH is narrow, with optimal performance occurring at or around a pH of 7.

Additional quantitative and graphing skills will be discussed at length in a later chapter of this text. There you will learn how to include statistics and hypothesis testing in your free-response questions and how best to present data using graphs.

UNDERSTAND CONTROL GROUPS IN EXPERIMENTS

Almost every experiment will have at least one group that represents the normal/unchanged/neutral version of the independent variable. This is called the **control group**. A control group is simply a standard of comparison. It enables the biologist to verify that the outcome of the study is due to changes in the independent variable and nothing else.

Let's say the principal of your school thinks that students who eat breakfast do better on the AP Biology Exam than those who don't eat breakfast. She gives a group of 10 students from your class free breakfast every day for a year. When the school year is over, she administers the AP Biology Exam, and they all score brilliantly! Did they do well because they ate breakfast every day? We don't know for sure. Maybe those kids were the smartest kids in the class and would have scored well anyway.

In this case, the best way to be sure that eating breakfast made a difference is to have a control group. In other words, the experiment would need to also follow for a year a group of students of equal intelligence to the first group BUT who are known to never eat breakfast. At the end of that year, both groups would take the AP Biology Exam, and if they performed similarly, you could probably conclude that eating breakfast wasn't necessarily leading to higher AP scores. The group of students who didn't eat breakfast is called the control group because those students were not "exposed" to the variable of interest—in this case, breakfast. Control groups are generally any group that you include just for the sake of comparison.

Put It All Together
Use the techniques discussed in this chapter, and practice writing your free responses. Additional sample free-response questions are available on apstudents.collegeboard.org/courses/ap-biology/assessment for the AP Biology Exam.

Chapter 3
Using Time Effectively to Maximize Points

BECOMING A BETTER TEST TAKER

Very few students stop to think about how to improve their test-taking skills. Most assume that if they study hard, they will test well and that if they do not study, they will do poorly. Most students continue to believe this even after experience teaches them otherwise. Have you ever studied really hard for an exam and then blown it on test day? Have you ever aced an exam for which you thought you weren't well prepared? Most students have had one, if not both, of these experiences. The lesson should be clear: factors other than your level of preparation influence your final test score. This chapter will provide you with some insights that will help you perform better on the AP Biology Exam and on other exams as well.

PACING AND TIMING

A big part of scoring well on an exam is working at a consistent pace. The worst mistake made by inexperienced or less savvy test-takers is that they come to a question that stumps them and rather than just skip it, they panic and stall. Time seems to stand still when you're working on a question you cannot answer, and it is not unusual for students to waste five minutes on a single question (especially a question involving a graph or the word EXCEPT) because they are too stubborn to cut their losses. It is important to be aware of how much time you have spent on a given question or section. There are several ways to improve your pacing and timing for the test:

- **Know your average pace.** While you prepare for your test, try to gauge how long you take on 5, 10, or 20 questions. Knowing how long you spend on average per question will help you identify how many questions you can answer effectively and how best to pace yourself for the test.

- **Have a clock-checking strategy.** The Bluebook testing app provides a countdown clock that you can view or hide. We recommend that you keep it hidden as your default, because constantly checking the clock is, in itself, a waste of time and can be distracting. Devise a plan. Try checking the clock after every 15 or 20 questions to see if you are keeping the correct pace or whether you need to speed up; this will ensure that you're cognizant of the time but will not permit you to fall into the trap of dwelling on it.

- **Know when to move on.** Since all questions are scored equally, investing too much time on a single question is inefficient and can potentially deprive you of the chance to answer easier ones later on. You should eliminate answer choices if you are able to, but don't feel bad about picking a random answer and moving on if you cannot find the correct answer. Remember, tests are like marathons: you do best when you work through them at a steady pace. You can always come back to a question you don't know. When you do, very often you will find that your previous mental block is gone and you will wonder why the question perplexed you the first time around (as you gleefully move on to the next question). Even if you still don't know the answer, you will not have wasted valuable time you could have spent on easier questions.

- **Be selective.** You don't have to do any of the questions in a given section in order. If you are stumped by an essay or multiple-choice question, skip it or choose a different one. In the next section, you will see that you may not have to answer every question correctly to achieve your desired score. Select the questions or essays that you can answer and work on them first. This will make you more efficient and give you the greatest chance of getting the most questions correct.

- **Use Process of Elimination (POE) on multiple-choice questions.** Many times, you can eliminate one or more answer choices. Every answer choice that you eliminate increases the odds that you will answer the question correctly.

Remember, when all the questions on a test are of equal value, no one question is that important, and your overall goal for pacing is to get the most questions correct. Finally, you should set a realistic goal for your final score.

GETTING THE SCORE YOU WANT

Depending on the score you need, it may be in your best interest not to try to work through every question. Check with the schools to which you are applying to find out what scores they will accept for credit.

Keep in mind, students are assessed only on the total number of correct answers. It's really important to remember that if you are running out of time, you should answer all the questions before the time for the multiple-choice section is up. Even if you don't plan to spend a lot of time on every question or even if you have no idea what the correct answer is, you need to fill something in.

Answer Every Question
Here's something worth repeating: there is no penalty for guessing. Don't leave a question blank; there's at least a 25% chance you might have gotten it right!

TEST ANXIETY

Everybody experiences anxiety before and during an exam. To a certain extent, test anxiety can be helpful. Some people find that they perform more quickly and efficiently under stress. If you've ever pulled an all-nighter to write a paper and ended up doing good work, you know the feeling.

However, too much stress is definitely a bad thing. Hyperventilating during the test, for example, almost always leads to a lower score. If you find that you stress out during exams, here are a few preemptive actions you can take:

- **Take a reality check.** Evaluate your situation before the test begins. If you have studied hard, remind yourself that you are well prepared. Remember that many others taking the test are not as well prepared, and (in your classes, at least) you are being graded against them, so you have an advantage. If you didn't study, accept the fact that you will probably not ace the test. Make sure you get to every question you know something about. Don't stress out or fixate on how much you don't know. Your job is to score as high as you can by maximizing the benefits of what you do know. In either scenario, it's best to think of a test as if it were a game. How can you get the most points in the time allotted to you? Always answer questions you can answer easily and quickly before tackling those that will take more time.

- **Try to relax.** Slow, deep breathing works for almost everyone. Close your eyes, take a few slow, deep breaths, and concentrate on nothing but your inhalation and exhalation for a few seconds. This is a basic form of meditation that should help you to clear your mind of stress and, as a result, concentrate better on the test. If you have ever taken yoga classes, you probably know some other good relaxation techniques. Use them when you can (obviously, anything that requires leaving your seat and, say, assuming a handstand position won't be allowed by any but the most free-spirited proctors).

- **Eliminate as many surprises as you can.** Make sure you know where the test will be given, when it starts, what type of questions are going to be asked, and how long the test will take. You don't want to be worrying about any of these things on test day or, even worse, after the test has already begun.

The best way to avoid stress is to study both the test material and the test itself. Congratulations! By reading this book, you are taking a major step toward a stress-free AP Biology Exam.

Part V
Content Review for the AP Biology Exam

4   Chemistry of Life
5   Cell Structure and Function
6   Cellular Energetics
7   Cell Communication and Cell Cycle
8   Heredity
9   Gene Expression and Regulation
10 Natural Selection
11 Ecology
12 Quantitative Skills and Biostatistics
13 Sample Free-Response Questions
14 Laboratory
15 Chapter Drill Answers and Explanations

REFLECT

Respond to the following questions:

- Which topics from this chapter do you feel you have mastered?

- Which content topics from this chapter do you feel you need to study more before you can answer multiple-choice questions correctly?

- Which content topics from this chapter do you feel you need to study more before you can effectively compose a free response?

- Was there any content that you need to ask your teacher or another person about?

Chapter 4
Chemistry of Life

ELEMENTS

Although life-forms exist in many diverse forms, they all have one thing in common: they are all made up of matter that is exchanged with the environment to grow, reproduce, and maintain organization. Matter is made up of elements. **Elements,** by definition, are substances that cannot be broken down into simpler substances by chemical means.

The Essential Elements of Life

Although there are 92 natural elements, 96% of the mass of all living things is made up of just four of them: **oxygen** (O), **carbon** (C), **hydrogen** (H), and **nitrogen** (N). These four elements are used to build biological molecules, such as carbohydrates, proteins, lipids, and nucleic acids. They are also used to form storage compounds and cells in all organisms. Other elements such as calcium (Ca), phosphorus (P), potassium (K), sulfur (S), sodium (Na), chlorine (Cl), and magnesium (Mg) are also present, but in smaller quantities. These elements make up most of the remaining four percent of a living thing's weight. Some elements are known as **trace elements** because they are required by an organism only in very small quantities. Trace elements include iron (Fe), iodine (I), and copper (Cu).

Where Are They Found?
Carbon and hydrogen and oxygen are found in all the macromolecules (essentially, very big molecules). Nitrogen is found in proteins and nucleic acids. Phosphorus is found in nucleic acids and some lipids.

SUBATOMIC PARTICLES

The smallest unit of an element that retains its characteristic properties is an **atom**. Atoms are the building blocks of the physical world.

Within atoms, there are even smaller subatomic particles called **protons**, **neutrons**, and **electrons**. Let's take a look at a typical atom.

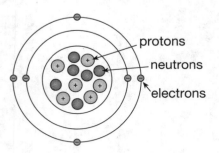

Protons and neutrons are particles that are packed together in the core of an atom called the **nucleus**. You'll notice that protons are positively charged (+) particles, whereas neutrons are uncharged particles.

Electrons, on the other hand, are negatively charged (−) particles that spin around the nucleus. Electrons are pretty small compared to protons and neutrons. In fact, for our purposes, electrons are considered massless. Most atoms have the same number of protons and electrons, making them electrically neutral. Some atoms have the same number of protons but differ in the number of neutrons in the nucleus. These atoms are called **isotopes**.

Carbon-14 Isotopes
Some isotopes are radioactive and decay predictably over time. Ancient artifacts can be dated by examining the rate of decay of carbon-14 and other isotopes within the artifact. This process is called **radiometric dating**.

COMPOUNDS

When two or more individual elements are combined in a fixed ratio, they form a chemical **compound**. You'll sometimes find that a compound has different properties from those of its elements. For instance, hydrogen and oxygen exist in nature as gases. Yet when they combine to make water, they often pass into a liquid state. When hydrogen atoms get together with oxygen atoms to form water, we've got a **chemical reaction**.

$$2H_2(g) + O_2(g) \rightarrow 2H_2O(l)$$

The atoms of a compound are held together by **chemical bonds**, which may be **ionic bonds**, **covalent bonds**, or **hydrogen bonds**.

An ionic bond is formed between two atoms when one or more electrons are transferred from one atom to the other. In order for this to occur, first, one atom *loses* electrons and becomes positively charged, and the other atom *gains* electrons and becomes negatively charged. The charged forms of the atoms are called **ions**. An ionic bond results from the attraction between the two oppositely charged ions. For example, when Na reacts with Cl, the charged ions Na$^+$ and Cl$^-$ are formed.

A covalent bond is formed when electrons are *shared* between atoms. If the electrons are shared equally between the atoms, the bond is called **nonpolar covalent**. If the electrons are shared unequally, the bond is called **polar covalent**. When one pair of electrons is shared between two atoms, the result is a single covalent bond. When two pairs of electrons are shared, the result is a double covalent bond. When three pairs of electrons are shared, the result is a triple covalent bond.

WATER: THE VERSATILE MOLECULE

One of the most important substances in nature is water. Water is considered a unique molecule because it plays an important role in chemical reactions.

Just like all biological molecules, the elements that make up water determine the properties of water. Water has two hydrogen atoms joined to an oxygen atom.

In water molecules, the electrons are not shared equally in the bonds between hydrogen and oxygen. This means that the hydrogen atoms have a partial positive charge, and the oxygen atom has a partial negative charge. Molecules that have partially positive and partially negative charges are said to be **polar**. Water is therefore a polar molecule. The positively charged elements of the water molecules strongly attract the negatively charged ends of other polar compounds (including water). Likewise, the negatively charged ends strongly attract the positively charged ends of polar compounds. These forces are most readily apparent in the tendency of water molecules to stick together, as in the formation of water beads or raindrops.

Water Weight
Did you know that more than 60 percent of your body weight consists of water?

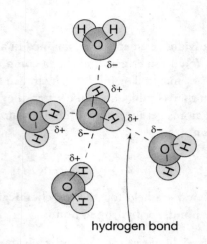
hydrogen bond

These types of intermolecular attractions are called hydrogen bonds. Hydrogen bonds are weak chemical bonds that form when a hydrogen atom that is covalently bonded to one electronegative atom is also attracted to another electronegative atom. Water molecules are held together by hydrogen bonds.

Although hydrogen bonds are individually weak, they are strong when present in large numbers. Because water reacts well with other polar substances, it makes a great solvent; it can dissolve many kinds of substances. The hydrogen bonds that hold water molecules together contribute to a number of special properties, including **cohesion**, **adhesion**, **surface tension**, **high heat capacity**, and **expansion on freezing**.

Cohesion and Adhesion

As mentioned above, water molecules have a strong tendency to stick together. That is, water exhibits cohesive forces. These forces are extremely important to life. For instance, during transpiration, water molecules evaporate from a leaf, "pulling" on neighboring water molecules. These, in turn, draw up the molecules immediately behind them, and so on, all the way down the plant vessels. The resulting chain of water molecules enables water to move up the stem.

Water molecules also like to stick to other substances—that is, they're **adhesive**. Have you ever tried to separate two glass slides stuck together by a film of water? They're difficult to separate because of the water sticking to the glass surfaces.

These two forces taken together—cohesion and adhesion—account for the ability of water to rise up the roots, trunks, and branches of trees. Since this phenomenon occurs in thin vessels, it's called **capillary action**.

Surface Tension

The cohesion of water molecules contributes to another property of water, its surface tension. Like a taut trampoline, the surface of water has a tension to it. The water molecules are stuck together and light things like leaves and water striders can sit atop the surface without sinking.

ACIDS AND BASES

We just said that water is important because most reactions occur in watery solutions. Well, there's one more thing to remember: reactions are also influenced by whether the solution in which they occur is **acidic**, **basic**, or **neutral**.

What makes a solution acidic or basic? A solution is acidic if it contains a lot of *hydrogen ions* (H^+). That is, if you dissolve an acid in water, it will release a lot of hydrogen ions. When you think about acids, you usually think of substances with a sour taste, like lemons. For example, if you squeeze a little lemon juice into a glass of water, the solution will become acidic. That's because lemons contain citric acid, which releases a lot of H^+ into the solution.

Bases, on the other hand, do not release hydrogen ions when added to water. They release a lot of *hydroxide ions* (OH^-). These solutions are said to be **alkaline** (the fancy name for a basic solution). Bases usually have a slippery consistency. Common soap, for example, is composed largely of bases.

The acidity or alkalinity of a solution can be measured with a **pH scale**. The pH scale is numbered from 1 to 14. The midpoint, 7, is considered neutral pH. The concentration of hydrogen ions in a solution will indicate whether it is acidic, basic, or neutral. If a solution contains a lot of hydrogen ions, then it will be acidic and have a low pH. Here's the trend:

> An increase in H^+ ions causes a decrease in the pH.
>
> $$pH = -\log [H^+]$$

Important Formulas
Don't forget to check out what formulas will be given on the Equations and Formulas sheet (pages 333–334).

One more thing to remember: the pH scale is not a linear scale—it's logarithmic. That is, a change of *one* pH number actually represents a *tenfold* change in hydrogen ion concentration. For example, a pH of 3 is actually ten times more acidic than a pH of 4. This is also true in the reverse direction: a pH of 4 represents a tenfold decrease in acidity compared to a pH of 3.

Therefore, as the concentration of H^+ ions increases by a factor of 10, the pH becomes one number smaller. For example, stomach acid has a pH between 1 and 3, and if we use the equation, we discover that the concentration of H^+ ions in stomach acid is 10^{-2} M. This is pretty high when you consider the other extreme is lye, which has a pH of 14, a concentration of H^+ ions around 10^{-14} M! Use your calculator to double-check that these numbers are correct.

You'll notice from the scale below that stronger acids have lower pHs. If a solution has a low concentration of hydrogen ions, it will have a high pH.

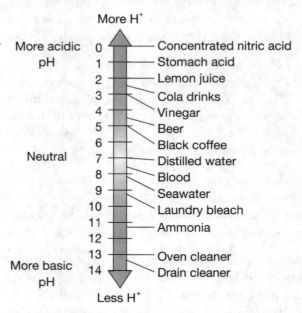

The equation for pH is listed on the AP Biology Equations and Formulas sheet. However, you will not be expected to perform calculations using this equation. Instead, you should understand how the equation works and when pH calculations are useful.

ORGANIC MOLECULES

Now that we've discussed chemical compounds in general, let's talk about a special group of compounds. Most of the chemical compounds in living organisms contain a skeleton of carbon atoms surrounded by hydrogen atoms and often other elements. These molecules are known as **organic compounds**. By contrast, molecules that do not contain carbon atoms are called **inorganic compounds**. For example, salt (NaCl) is an inorganic compound.

Carbon is important for life because it is a versatile atom, meaning that it has the ability to bind not only with other carbons but also with a number of other elements including nitrogen, oxygen, and hydrogen. The resulting molecules are key in carrying out the activities necessary for life. Most macromolecules are chains of building blocks, called **polymers**. The individual building blocks of a polymer are called **monomers**.

Dehydration Synthesis and Hydrolysis Reactions

Polymers are formed through **dehydration synthesis** (or **condensation**) reactions. In these reactions, a water molecule is lost, and a larger compound is formed. Polymers can also be broken down into monomers. That's called **hydrolysis**, which means "water" (hydro-) and "breaking" (-lysis). The water breaks the bond between the two monomers.

Break Down the Words
Dehydration = removal of water
Synthesis = Creation
Hydro = water
Lysis = breakdown

Dehydration Synthesis

Hydrolysis (Digestion)

Look for This Theme!
Dehydration synthesis is a common reaction for joining things, and hydrolysis is a common reaction for separating things.

To recap:

- Organic compounds contain carbon atoms AND hydrogen atoms (and sometimes other things too).
- Inorganic compounds don't contain both carbon atoms and hydrogen atoms.
- Polymers are formed through dehydration synthesis reactions and broken down through hydrolysis reactions.

Now let's focus on the following four classes of organic compounds central to life on Earth:

- carbohydrates
- proteins
- lipids
- nucleic acids

A Structured Approach
While it's a good idea to be familiar with the variety of molecules and to know how they differ, note that for the AP Exam, you will not have to memorize the structures of specific nucleotides, amino acids, proteins, lipids, or carbohydrates.

Carbohydrates

Carbohydrates are organic compounds that contain carbon, hydrogen, and oxygen. They usually contain these three elements in a ratio of approximately 1:2:1, respectively. We can represent the proportion of elements within carbohydrate molecules by the formula $C_nH_{2n}O_n$, which can be simplified as $(CH_2O)_n$.

Most carbohydrates are categorized as either **monosaccharides**, **disaccharides**, or **polysaccharides**. The term *saccharides* is a fancy word for "sugar." The prefixes *mono-*, *di-*, and *poly-* refer to the number of sugars in the molecule. *Mono-* means "one," *di-* means "two," and *poly-* means "many." A monosaccharide is therefore a carbohydrate made up of a single saccharide.

Monosaccharides: The Simplest Sugars

Monosaccharides, the simplest sugars, serve as an energy source for cells. The two most common sugars are (1) **glucose** and (2) **fructose**. Galactose, ribose, and deoxyribose are also monosaccharides.

Both glucose and fructose are six-carbon sugars with the chemical formula $C_6H_{12}O_6$. Glucose, the most abundant monosaccharide, is the most popular sugar around. Glucose is an important part of the food we eat, and it is the favorite food made by plants during photosynthesis. Living organisms break down glucose in order to provide cells with energy. Almost all living organisms can perform this ancient biochemical process (which is called cell respiration; you'll learn more about this process in Chapter 6). In fact, the evolution of glycolysis, photosynthesis, and oxidative metabolism is thought to have occurred about 3 billion years ago! Fructose, the other monosaccharide you need to know for the test, is a common sugar in fruits.

Glucose and fructose can be depicted as either "straight" or "rings." Both of them are pretty easy to spot; just look for the carbon molecules with a lot of OHs and Hs attached to them.

Here are the two different forms.

Ring form of glucose

Straight-chain form of glucose

Ring form of fructose

Straight-chain form of fructose

Disaccharides

What happens when two monosaccharides are brought together? The hydrogen (–H) from one sugar molecule combines with the hydroxyl group (–OH) of another sugar molecule. What do H and OH add up to? Water (H_2O)!

When two monosaccharides are joined, the bond is called a **glycosidic linkage,** and the resulting sugar is called a disaccharide. The disaccharide formed from two glucose molecules is maltose. Other common disaccharides include sucrose (table sugar) and lactose (found in dairy products).

Now, what if you want to break up the disaccharide and form two monosaccharides again? Just add water.

Digest This Tidbit!
Did you ever look at a pile of firewood and think, Yummy!? Probably not. Wood is almost entirely made of cellulose, which contains β-linked glucose that humans can't digest. When we eat plants, we can get nutrients from their stored starch, but not from the cellulose of their cell walls. This passes through us as fiber roughage. Fiber is important in the diet for other reasons, but you should probably leave the firewood alone.

Polysaccharides

Polysaccharides are made up of many repeated units of monosaccharides. Polysaccharides can consist of branched or unbranched chains of monosaccharides. The most common polysaccharides you'll need to know for the test are **starch**, **cellulose**, and **glycogen**. Glycogen and starch are sugar storage molecules. Glycogen stores sugar in animals and starch stores sugar in plants. Cellulose, on the other hand, is made up of β-glucose and is a major part of the cell walls in plants. Its function is to lend structural support. Chitin, a polymer of β-glucose molecules, serves as a structural molecule in the walls of fungus and in the exoskeletons of arthropods.

To illustrate the concept of polysaccharide branching, take a look below at two components of starch: amylose and amylopectin. Amylose is an unbranched polysaccharide and is thus linear, whereas amylopectin has many branched chains of glucose monomers.

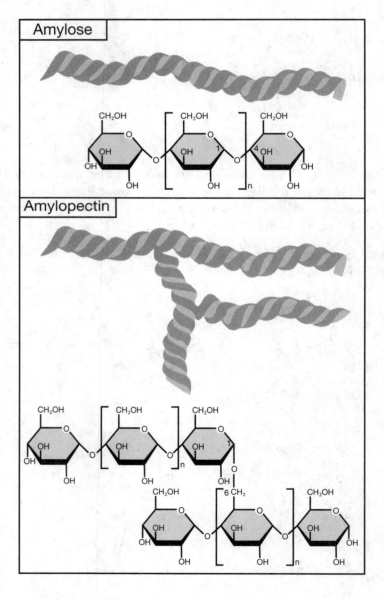

Proteins

Proteins perform most of the work in your cells, and are important for structure, function, and regulation of your tissues and organs. There are thousands of different types of proteins. **Amino acids** are organic molecules that serve as the building blocks of proteins. They contain carbon, hydrogen, oxygen, and nitrogen atoms. There are 20 different amino acids commonly found in proteins. Fortunately, you don't have to memorize all 20 amino acids. However, you do have to remember that every amino acid has four important parts around a central carbon: an **amino group** ($-NH_2$), a **carboxyl group** ($-COOH$), a hydrogen, and an **R-group**.

Here's a typical amino acid.

Amino acids differ only in the R-group, which is also called the **side chain**. The R-group associated with an amino acid could be as simple as a hydrogen atom (as in the amino acid *glycine*) or as complex as a charged carbon skeleton (as in the amino acid *arginine*).

Glycine

Arginine

When it comes to spotting an amino acid, simply keep an eye out for the amino group (NH_2); then look for the carboxyl molecule ($COOH$). The side chains for each amino acid vary greatly. You don't have to memorize the structures of the side chains for individual amino acids, but you should know that they can vary in composition, polarity, charge, and shape depending on the side chain that they have.

How do the side chains of the amino acids differ?

- Composition of elements (C, H, O, N, and S)
- Polarity (polar, nonpolar)
- Charge (neutral, positive, negative)
- Shape (long chain, short chain, ring shape)

Check out your online Student Tools for a reference page of the 20 naturally occurring amino acids. You won't have to memorize their structures for the test, but it can be helpful to familiarize yourself with them.

Building a Chain
The new amino acid is always added on the carboxyl end of the existing chain.

Side chain polarity affects whether an amino acid is more hydrophobic or more hydrophilic. While there are different ways of categorizing amino acids, the AP Exam divides them into 3 broad categories: **hydrophobic** (non-polar and uncharged), **hydrophilic** (polar and uncharged), and **ionic** (polar and charged). Of the common amino acids, two (glutamic acid and aspartic acid) donate a proton, making them negatively charged. Two (lysine and arginine) accept a proton at physiological pH, which makes them positively charged. Finally, two contain the atom sulfur: methionine and cysteine.

Polypeptides

When two amino acids join, they form a **dipeptide**. The carboxyl group of one amino acid combines with the amino group of another amino acid. Here's an example:

Here's the peptide bond

This is the same dehydration synthesis process we saw earlier. Why? Because a water molecule is removed to form a bond. By the way, the bond between two amino acids has a special name—a **peptide bond**. If a group of amino acids is joined together in a "string," the resulting organic compound is called a **polypeptide**. Once a polypeptide chain twists and folds on itself, it forms a three-dimensional structure called a **protein**.

glycine residue   alanine residue   alanine residue   glycine residue

glycylalanine (Gly-Ala)   alanylglycine (Ala-Gly)

The specific order of amino acids in a peptide can change the identity of the peptide and determine how the peptide affects the overall shape of a protein. In the dipeptide above, notice that one end of the peptide has an amino group (NH_2) while one end has a carboxyl group (COOH).

Generally, all peptides will have an end with an amino group, called the N-terminus or amino terminus, in addition to an end with a carboxyl group, called C-terminus or carboxyl terminus. Based on the specific way that two amino acids bond, two different dipeptides can result. For example, both dipeptides above have the same amino acids. However, one has glycine at the N-terminus and alanine at the C-terminus, whereas the other has alanine at the N-terminus and glycine at the C-terminus.

Higher Protein Structure

The polypeptide has to go through several changes before it can officially be called a protein. Proteins can have four levels of structure. The linear sequence of the amino acids is called the **primary structure** of a protein. Now the polypeptide begins to twist, forming either a coil (known as an alpha helix) or zigzagging pattern (known as beta-pleated sheets). These are examples of proteins' **secondary structures**.

A polypeptide folds and twists because the different R-groups of the amino acids are interacting with each other. Remember, each R-group is unique and has a particular size, shape, charge, and so on, that allows it to react to things around it. Depending on which amino acids are in a protein and the order that they are in, the protein can twist and fold in very different ways. This is why proteins can form so many different shapes.

The secondary structure is formed by amino acids that interact with other amino acids close by in the primary structure. However, after the secondary structure reshapes the polypeptide, amino acids that were far away in the primary structure arrangement can now also interact with each other. This is called the **tertiary structure**. Because most proteins are found in an aqueous environment, hydrophilic amino acids and regions of the peptide chain are often located on the exterior of the protein. Hydrophobic amino acids and regions are usually found on the interior of proteins. Sometimes, two cysteine amino acids can react with each other to form a covalent disulfide bond that stabilizes the tertiary structure. Tertiary structure often minimizes the free energy of the molecule and locks it into a stable 3D shape.

Lastly, several different polypeptide chains sometimes interact with each other to form a **quaternary structure**. The different polypeptide chains that come together are often called subunits of the final whole protein.

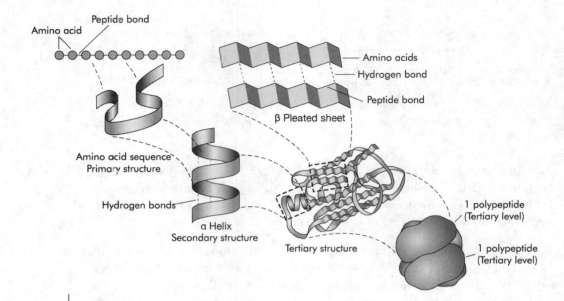

A classic example of a protein with quaternary structure is hemoglobin. Hemoglobin is a molecule in the blood that helps distribute oxygen to the tissues in the body. It is formed when four separate polypeptide chains interact with each other. Each of the four polypeptide chains creates a subunit of the final hemoglobin protein.

Hemoglobin

Only proteins that have folded correctly into their specific three-dimensional structure can perform their function properly. Mistakes in the amino acid chain can create oddly shaped proteins that are nonfunctional.

One more thing: in some cases, the folding of proteins involves other proteins known as **chaperone proteins** (**chaperonins**). They help the protein fold properly and make the process more efficient.

Lipids

Like carbohydrates, **lipids** consist of carbon, hydrogen, and oxygen atoms, but not in the 1:2:1 ratio typical of carbohydrates. The most common examples of lipids are **triglycerides, phospholipids**, and **steroids.** Lipids are important because, due to their nonpolar structures, they function as structural components of cell membranes, sources of insulation, signaling molecules, and a means of energy storage.

Triglycerides

Our bodies store fat in tissue called adipose, which is made of cells called adipocytes; these cells are filled with lipids called triglycerides. Each triglyceride is made of a glycerol molecule (sometimes called the glycerol backbone) with three fatty acid chains attached to it. A fatty acid chain is mostly a long chain of carbons in which each carbon is covered in hydrogen. One end of the chain has a carboxyl group (just like we saw in an amino acid).

A fatty acid can be **saturated** with hydrogens along its long carbon chain, or it can have a few gaps where double bonds exist instead of a hydrogen. If there is a double bond in the chain, then it is an **unsaturated** fatty acid. A **polyunsaturated** fatty acid has many double bonds within the fatty acid.

Lipid Saturation

The extent of saturation in a lipid can affect its structure and function. The more double bonds that exist within a lipid, the more unsaturated it is:

Notice the difference above between the saturated and unsaturated fatty acid. In general, saturated fatty acids are relatively linear molecules, whereas unsaturated fatty acids with *cis*-double bonds, as above, are more kinked in their structure. When molecules are linear, they are more easily able to stack together and interact with each other more tightly; thus saturated fats tend to form solids at room temperature (an example is butter). When molecules are kinked, they cannot pack as tightly and tend to be liquid at room temperature (an example is an oil).

> **Make the Connection**
> Amino acids have carboxyl groups and fatty acids have carboxyl groups. Carboxyl groups are acidic; hence they both have acid in their name.

One exception to this rule is an unsaturated fatty acid with a *trans*-double bond, as below:

trans-Oleic acid

cis-Oleic acid

With this trans-double bond, the unsaturated lipid is relatively linear, as would be expected in a saturated fat. Due to the relative linearity, they are more likely than unsaturated lipids with *cis*-double bonds to be a solid at room temperature.

Phospholipids

Another special class of lipids is known as phospholipids. Phospholipids contain two fatty acid "tails" and one negatively charged phosphate "head." Take a look at a typical phospholipid.

Phospholipids are extremely important, mainly because of some unique properties they possess, particularly with regard to water.

The two fatty acid tails are hydrophobic ("water-hating"). In other words, just like oil and vinegar, fatty acids and water don't mix. The reason for this is that fatty acid tails are nonpolar, and nonpolar substances don't mix well with polar ones, such as water.

On the other hand, the phosphate "head" of the lipid is hydrophilic ("water-loving"), meaning that it does mix well with water. Why? It carries a negative charge, and this charge draws it to the positively charged end of a water molecule. Since a phospholipid has both a hydrophilic region and a hydrophobic region, it is an **amphipathic molecule**. One side of a phospholipid loves to hang out with water, and the other side hates to.

Thus, the two fatty acid chains orient themselves away from water, while the phosphate portion orients itself toward the water. Keep these properties in mind. We'll see later how this orientation of phospholipids in water relates to the structure and function of cell membranes.

Cholesterol

Cholesterol is another important type of lipid. Cholesterol is a four-ringed molecule that is found here and there in membranes. It generally increases membrane fluidity, except at very high temperatures when it helps to hold things together instead. Cholesterol is also important for making certain types of hormones and for making vitamin D. Here are some examples:

Estradiol

Testosterone

Cholesterol

Nucleic Acids

The fourth class of organic compounds is the **nucleic acids.** Like proteins, nucleic acids contain carbon, hydrogen, oxygen, and nitrogen, but nucleic acids also contain phosphorus. Nucleic acids are molecules that are made up of simple units called **nucleotides.** For the AP Biology Exam, you'll need to know about two kinds of nucleic acids: **deoxyribonucleic acid (DNA)** and **ribonucleic acid (RNA).**

Need-to-Know Nucleic Acids

For the AP Biology Exam, you'll be expected to know about two types of nucleic acids, DNA and RNA, which are shown here.

DNA is important because it contains the hereditary "blueprints" of all life. RNA is important because it's essential for protein synthesis. We'll discuss DNA and RNA in greater detail when we discuss heredity (see Chapter 8). The following table summarizes the important macromolecules discussed here.

Important Biological Macromolecules

Macromolecule	Monomer	Polymer	Linkage Bond	Elements
Carbohydrates	Monosaccharide (e.g., Glucose)	Polysaccharide (e.g., Starch, glycogen, cellulose)	Glycosidic linkage	C, H, O,
Proteins	Amino Acid (e.g., Glycine)	Polypeptide (e.g., Actin)	Peptide bond	C, H, O, N, S
Nucleic Acids	Nucleotides (e.g., Adenine, thymine, guanine, cytosine)	Deoxyribonucleic acid (DNA) Ribonucleic acid (RNA)	Sugar-phosphate phosphodiester bonds	C, H, O, N, P
Lipids	Not a true polymer, but often contains chains of carbons with hydrogens	Triglycerides, phospholipids, cholesterol	Ester bonds	C, H, O, sometimes P

CHAPTER 4 KEY TERMS

elements
oxygen
carbon
hydrogen
nitrogen
trace elements
atom
protons
neutrons
electrons
nucleus
isotopes
radiometric dating
compound
chemical reaction
chemical bond
ionic bond
covalent bond
hydrogen bond
ions
nonpolar covalent
polar covalent
polar
cohesion
adhesion
surface tension
high heat capacity
expansion on freezing
adhesive
capillary action
glycerol
acidic
basic
neutral
alkaline
pH scale
organic compounds
inorganic compounds
polymer
monomer
dehydration synthesis
condensation

hydrolysis
carbohydrates
monosaccharides
disaccharides
polysaccharides
glucose
fructose
glycosidic linkage
starch
cellulose
glycogen
amino acids
amino group
carboxyl group
R-group
side chain
hydrophobic
hydrophilic
ionic
dipeptide
peptide bond
polypeptide
protein
primary structure
secondary structure
tertiary structure
quaternary structure
chaperone proteins (chaperonins)
lipids
triglycerides
phospholipids
steroids
ester linkage
saturated
unsaturated
polyunsaturated
amphipathic molecule
cholesterol
nucleic acids
nucleotides
deoxyribonucleic acid (DNA)
ribonucleic acid (RNA)

Chapter 4 Drill

Answers and explanations can be found in Chapter 15.

1. Mark for Review

Water is a critical component of life due to its unique structural and chemical properties. Which of the following describes a way that the exceptional characteristics of water are used in nature to sustain life?

(A) The low polarity of water prevents dissolution of cells and compounds.

(B) The low surface tension and cohesiveness of water facilitates capillary action in plants.

(C) The high heat capacity of water prevents lakes and streams from rapidly changing temperature and freezing completely solid in the winter.

(D) The low intermolecular forces of water, such as hydrogen bonding, result in a boiling point which exceeds the tolerance of most life on the planet.

2. Mark for Review

Biomolecules are held together by intramolecular forces between atoms. Biomolecules also interact with other biomolecules through intermolecular forces, such as hydrogen bonds. Which of the following interactions is an example of a hydrogen bond?

(A) The bond between a carbon atom and an oxygen atom in propanol.

(B) The bond between a carbon atom and an oxygen atom in a carbonyl group in a ketone.

(C) The attraction between the hydrogen of one water molecule and the oxygen of another water molecule.

(D) The glycosidic bond between sugar molecules in a polysaccharide.

3. Mark for Review

Proteins are polymers made up of monomers of amino acids. Which of the following functional groups make up all amino acids?

(A) Carboxyl group and amino group

(B) Hydroxyl group and amino group

(C) Carboxyl group and aldehyde group

(D) Hydroxyl group and aldehyde group

4. Mark for Review

The chemical structure for the nucleic acid RNA is shown above. It is an important biomolecule responsible for creating proteins via transcription and translation. Which of the following are its monomers?

(A) Glycerol and fatty acids

(B) Monosaccharides

(C) Amino acids

(D) Nucleotides

5 Mark for Review

Adenosine triphosphate, known as ATP, is made up of a ribose sugar, an adenine nucleotide, and three negatively charged phosphate groups. The phosphate groups are easily detached from the ATP molecule. Which of the following best explains why ATP is utilized as the primary energy source for most organisms?

(A) Remnants of ATP have been found in specimens dating back to the origination of life on Earth.

(B) ATP is the smallest unit of energy inside living organisms.

(C) ATP is stable, so living organisms can store it to use in the future.

(D) The components of ATP can be easily broken down to perform functions for living organisms.

6 Mark for Review

Hydrochloric acid (HCl), which has a pH of approximately 1.5, is one of the primary contents of the human stomach. The acid helps activate digestive enzymes to efficiently digest food. The stomach sheds its lining approximately once every three days, but if the acid concentration is too high, humans can experience swelling, bleeding, and ulcers. Some $Mg(OH)_2$ antacids contain a buffer to temporarily neutralize excess HCl in the stomach. Which of the following methods for the neutralization of HCl provides a buffering system for the stomach?

(A) The hydroxide (OH^{-1}) ions combine with excess hydrogen (H^{+1}) ions ions to form water molecules and are transported out of the body as urine.

(B) The magnesium (Mg^{+2}) ions combine with excess hydrogen (H^{+1}) ions and are transported out of the body as feces.

(C) The $Mg(OH)_2$ antacid combines with water molecules to dilute the concentration of hydrogen (H^{+1}) ions.

(D) $Mg(OH)_2$ is converted into $Mg(OH)_3$ to absorb the excess hydrogen (H^{+1}) ions.

7 Mark for Review

Anabolic reactions require energy to build macromolecules from monomers. The synthesis of sucrose requires the combination of one molecule of glucose and one molecule of fructose. What is always produced as a byproduct of this reaction?

(A) Ammonia

(B) Water

(C) Carbon dioxide

(D) Galactose

8 Mark for Review

The fungus *Candida parapsilosis* can cause infection and sepsis when tissue is damaged in living organisms. A new drug is being developed that can target the alpha helices and beta-pleated sheets of proteins that are essential for the fungus but not present in vertebrates, such as humans. Which level of protein structure does the drug target?

(A) Primary structure

(B) Secondary structure

(C) Tertiary structure

(D) Quaternary structure

9 ⬜ Mark for Review

Many marine mammals and fish live in the Arctic Ocean under extremely harsh conditions. The North Polar region is home to many species of walruses, seals, dolphins, and whales. Which of the following is used to provide buoyancy and insulation to these species of marine Arctic organisms?

- Ⓐ Carbohydrates
- Ⓑ Nucleic acids
- Ⓒ Lipids
- Ⓓ Proteins

10 ⬜ Mark for Review

Cholesterol is a lipid composed of four rings found sporadically in cell membranes. It increases the fluidity of cell membranes by inserting into the phospholipid's fatty acid tails and preventing them from aggregating together. Which of the following properties accurately describes cholesterol?

- Ⓐ Cholesterol maintains cell membrane fluidity at all temperatures.
- Ⓑ Cholesterol is soluble in water.
- Ⓒ Cholesterol is a hydrophobic molecule.
- Ⓓ Cholesterol has no benefits for humans and causes numerous health issues.

Chapter 4 Summary

- Compounds are composed of two or more types of elements in a fixed ratio.

- Molecules are held together by ionic bonds and covalent bonds.

- Hydrogen bonding is essential for the following emergent properties of water:
  - cohesion
  - adhesion
  - surface tension
  - high heat capacity
  - expansion on freezing
  - polar structure
  - ability to dissolve other polar (hydrophilic) substances but inability to dissolve nonpolar (hydrophobic) substances

- pH is important for biological reactions. Solutions may be acidic, basic or alkaline, or neutral.

- Organic molecules contain carbon atoms. Many molecules in biology are composed of several monomers bound together into polymers. The creation of these polymers occurs through dehydration synthesis reactions. The breakdown of these polymers occurs through hydrolysis reactions.

- Important organic macromolecules in biology include:
  - Carbohydrates: monosaccharides link by glycosidic bonds to form di- and polysaccharides. Examples are glucose, fructose, glycogen, cellulose, and starch.
  - Proteins: 20 different amino acids, each with a different R-group, link together via peptide bonds to form polypeptides. The chain further folds up into special shapes to make the final protein structure. The structure can be divided into primary (amino acid sequence), secondary (alpha helices and beta-pleated sheets), tertiary (disulfide bonds), and quaternary (multiple polypeptides).
  - Lipids: chains called fatty acids (either saturated or unsaturated) combine on a glycerol backbone to form phospholipids and triglycerides. Cholesterol is another type of lipid. Lipids are hydrophobic.
  - Nucleic acids: chains of either DNA or RNA nucleotides. They carry the genetic recipes in the body.

Chapter 5
Cell Structure and Function

LIVING THINGS

All living things are composed of **cells**. According to the cell theory, the cell is life's basic unit of structure and function. This simply means that the cell is the smallest unit of living material that can carry out all the activities necessary for life.

It may seem strange that we are still made of tiny little cells. Why not just be a living thing that is one giant cell?

One reason for this is because compartments allow more specialization. As we will see, compartmentalization is an important part of the organization of unicellular and multicellular organisms.

The second reason we can't just be giant cells is the **surface area-to-volume ratio**. There is always lots of exchange going on between the inside of things and the outside. This ratio must be kept large so that there is plenty of space to do these exchanges. Some parts of living things are special because they have many folds, which increase the surface area. Generally, smaller cells have a higher surface area-to-volume ratio; this allows them to more efficiently exchange materials with their surroundings through specialized exchange surfaces. As cells increase in volume, the surface area-to-volume ratio decreases, and the exchange of materials becomes less efficient.

The surface area and volumes of cells can be calculated using typical geometry formulas listed below. Different types of cells have different shapes, so be careful to use the appropriate formula if you're asked to calculate the ratio!

> **Surface Area-to-Volume Ratio**
>
> This is a concept that comes up often. Use the volume formulas on the Equations and Formulas sheet to practice calculating the surface area-to-volume ratios of different-sized cubes and spheres.

Relevant Equations

Volume of a Sphere: $V = \dfrac{4}{3}\pi r^3$

Volume of a Cube: $V = s^3$

Volume of a Rectangular Solid:
$V = lwh$

Volume of a Cylinder: $V = \pi r^2 h$

Surface Area of a Sphere: $SA = 4\pi r^2$

Surface Area of a Cube: $SA = 6s^2$

Surface Area of a Rectangular Solid:
$SA = 2lh + 2lw + 2wh$

Surface Area of a Cylinder:
$SA = 2\pi rh + 2\pi r^2$

r = radius

l = length

h = height

w = width

s = length of one side of a cube

The surface area-to-volume ratio concept can also be applied to organisms. As organisms increase in size, their ratio will decrease, and this can affect properties of the organism, such as heat exchange with their surroundings. Small organisms lose heat at much higher rates than larger organisms as a result of their efficient exchange of heat.

Remember, even the largest things are still made of cells.

TYPES OF CELLS AND ORGANELLES

For centuries, scientists have known about cells. However, it wasn't until the development of the electron microscope that scientists were able to figure out what cells do. We now know that there are two distinct types of cells: **prokaryotic cells** and **eukaryotic cells**.

A prokaryotic cell, which is a lot smaller than a eukaryotic cell, is relatively simple. Bacteria and archaea are examples of prokaryotes. The inside of the cell is filled with a substance called **cytoplasm**. The genetic material in a prokaryote is one continuous, circular DNA molecule that is found free in the cell in an area called the **nucleoid** (this is not the same as a nucleus!). Most prokaryotes have a **cell wall** composed of peptidoglycans that surrounds a lipid layer called the **plasma membrane**. Prokaryotes also have ribosomes (though smaller than those found in eukaryotic cells). Some bacteria may also have one or more **flagella**, which are long projections used for motility (movement) and they might have a thick **capsule** outside their cell wall to give them extra protection.

Eukaryotic cells are more complex than prokaryotic cells. Fungi, protists, plants, and animals are eukaryotes. Eukaryotic cells are organized into many smaller structures called **organelles**. Some of these organelles are the same structures seen in prokaryotic cells, but many are uniquely eukaryotic. A good way to remember the difference is that prokaryotes do not have any membrane-bound organelles. Their only membrane is the plasma membrane.

Cells are studied with different types of microscopes. **Light microscopes** (for example, the compound microscopes commonly found in labs) are used to study stained or living cells. They can magnify the size of an organism up to 1,000 times. **Electron microscopes** are used to study detailed structures of a cell that cannot be easily seen or observed by light microscopy. They are capable of resolving structures as small as a few nanometers in length, such as individual virus particles or the pores on the surface of the nucleus.

Endosymbiosis
It is believed that chloroplasts and mitochondria used to be independent prokaryotic cells. They have since become a permanent part of eukaryotic cells.

Organelles

A eukaryotic cell is like a microscopic factory. It's filled with organelles, each of which has its own special tasks. Let's take a tour of a eukaryotic cell and focus on the structure and function of each organelle. Here's a picture of a typical animal cell and its principal organelles.

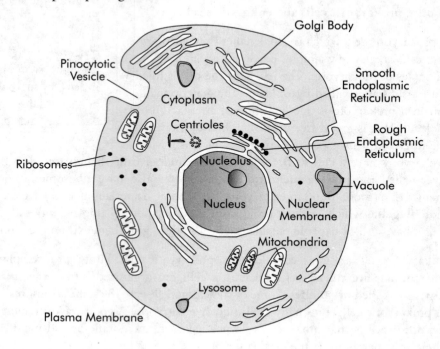

Plasma Membrane

The cell has an outer envelope known as the plasma membrane. Although the plasma membrane appears to be a simple thin layer surrounding the cell, it's actually a complex, double-layered structure made up of mostly phospholipids and proteins. The *hydrophobic* fatty acid tails face inward and the *hydrophilic* phosphate heads face outward. It is called a **phospholipid bilayer** since two lipid layers are forming a hydrophobic sandwich.

The plasma membrane is important because it regulates the movement of substances into and out of the cell. The membrane itself is semipermeable, meaning that only certain substances, namely small hydrophobic molecules (such as O_2 and CO_2), pass through it unaided. Anything large and/or hydrophilic can pass through the membrane only via special tunnels (discussed later in this chapter).

Many proteins are associated with the cell membrane. Some of these proteins are loosely associated with the lipid bilayer (**peripheral proteins**). They are located on the inner or outer surface of the membrane. Others are firmly bound to the plasma membrane (**integral proteins**). These proteins are amphipathic, which means that their hydrophilic regions extend out of the cell or into the cytoplasm, while their hydrophobic regions interact with the tails of the membrane phospholipids. Some integral proteins extend all the way through the membrane (**transmembrane proteins**).

> The surface area of a cell must be large enough to allow sufficient exchange of materials. In general, smaller cells have a more favorable surface area-to-volume ratio.

This arrangement of phospholipids and proteins is known as the **fluid-mosaic model**. This means that each layer of phospholipids is flexible, and it is a mosaic because it is peppered with different proteins and carbohydrate chains. Remember, anything hydrophilic should not go through the hydrophobic interior. This means that the phospholipids on one side should never flip-flop to the other side of the membrane (because that would require their polar heads to pass through the hydrophobic area).

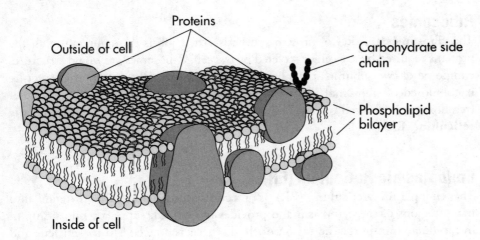

Why should the plasma membrane need so many different proteins? It's because of the number of activities that take place in or on the membrane. Generally, plasma membrane proteins fall into several broad functional groups. Some membrane proteins (**adhesion proteins**) form junctions between adjacent cells. Others (**receptor proteins**), such as hormones, serve as docking sites for arrivals at the cell. Some proteins (**transport proteins**) form pumps that use ATP to actively transport solutes across the membrane. Others (**channel proteins**) form channels that selectively allow the passage of certain ions or molecules. Finally, some proteins (**cell surface markers**), such as glycoproteins, and some lipids, such as glycolipids, are exposed on the extracellular surface and play a role in cell recognition and adhesion.

Attached to the surface of some proteins are **carbohydrate side chains**. They are found only on the outer surface of the plasma membrane. As mentioned above, cholesterol molecules are also found in the phospholipid bilayer because they help stabilize membrane fluidity in animal cells.

Ribosomes are found in both eukaryotes and prokaryotes. Eukaryotic ribosomes (80s) are slightly larger than prokaryotic ribosomes (70s). The "s" is a unit of measurement used to compare sizes of small things. In general, prokaryotic ribosomes are exclusively found freely in the cytoplasm, whereas eukaryotic ribosomes can be found freely in the cytoplasm and attached to the endoplasmic reticulum. The fact that both eukaryotes and prokaryotes possess ribosomes underlines the common ancestry of all known life.

The flattened membrane disks of the endoplasmic reticulum and Golgi apparatus are called cisternae. A Golgi stack may contain anywhere from 3 to 20 cisternae, and these are divided into cis (close to the RER), medial (in the middle), and trans (farthest from the RER). Each cisternae type contains different types of enzymes, and molecules can move between cisternae via vesicles.

More Mitochondria = More Energy

Since mitochondria are the cell's powerhouses, you're likely to find more of them in cells that require a lot of energy. Muscle cells, for example, are rich in mitochondria.

The Nucleus

The **nucleus** is usually the largest organelle in the cell. The nucleus not only directs what goes on in the cell, but is also responsible for the cell's ability to reproduce. It's the home of the hereditary information—DNA—which is organized into large structures called **chromosomes**. The most visible structure within the nucleus is the **nucleolus**, which is where rRNA is made and ribosomes are assembled.

Ribosomes

The **ribosomes** are sites of protein synthesis. Their job is to manufacture all the proteins required by the cell or secreted by the cell. Ribosomes are round structures composed of two subunits, the large subunit and the small subunit. The structure is composed of ribosomal RNA (rRNA) and proteins. Ribosomes can be either free floating in the cell or attached to another structure called the **endoplasmic reticulum (ER)**.

Endoplasmic Reticulum (ER)

The endoplasmic reticulum (ER) is a continuous channel that extends into many regions of the cytoplasm and provides mechanical support while aiding in intracellular transport. The region of the ER that is attached to the nucleus and "studded" with ribosomes is called the **rough ER** (RER), which compartmentalizes the cell. Proteins generated in the rough ER are trafficked to or across the plasma membrane, or they are used to build Golgi bodies, lysosomes, or the ER. The region of the ER that lacks ribosomes is called the **smooth ER** (SER). The smooth ER makes lipids, hormones, and steroids and breaks down toxic chemicals.

Golgi Complex

The **Golgi complex**, which looks like stacks of flattened sacs, also participates in the processing of proteins. Once the ribosomes on the rough ER have completed synthesizing proteins, the Golgi complex modifies, processes, and sorts the products. It is the packaging and distribution center for materials destined to be sent out of the cell. It packages the final products in little sacs called **vesicles**, which carry products to the plasma membrane. The Golgi complex is also involved in the production of lysosomes.

Mitochondria

Another important organelle is the mitochondrion. **Mitochondria** are often referred to as the "powerhouses" of the cell. They're power stations responsible for converting energy from organic molecules into useful energy for the cell. The most common energy molecule in the cell is **adenosine triphosphate (ATP)**.

The mitochondrion is usually an easy organelle to recognize because it has a unique oblong shape and a characteristic double membrane consisting of an inner portion and an outer portion. The inner mitochondrial membrane forms folds, known as

cristae, and separates the innermost area (the matrix) from the intermembrane space. The outer membrane separates the intermembrane space from the cytoplasm. As we'll see later, most of the production of ATP is done on the cristae. Having folds in the membrane increases the surface area for making ATP.

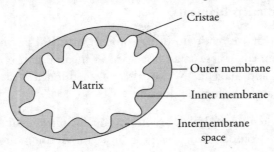

> There are lots of examples in biology in which folding is used to increase surface area. This is similar to the surface area-to-volume ratio concept we mentioned earlier. Some common examples are inner mitochondrial membrane, ER membrane, convoluted membranes in chloroplasts, brain tissue, small intestine (villi and microvilli), alveoli in the lungs, or root hairs in plants.

Lysosomes

Throughout the cell are small, membrane-enclosed structures called **lysosomes**. These tiny sacs carry digestive enzymes, which they use to break down old, worn-out organelles, debris, or large ingested particles. Lysosomes make up the cell's clean-up crew, helping to keep the cytoplasm clear of unwanted flotsam and recycling organic materials. Lysosomes contain hydrolytic enzymes that function only at acidic pH, which is enclosed inside the lumen of the lysosome. Lysosomes are made when vesicles containing specific enzymes from the trans Golgi fuse with vesicles made during endocytosis (which you'll learn about in a bit). Lysosomes are also essential during programmed cell death, called apoptosis.

Centrioles

The **centrioles** are small, paired, cylindrical structures that are often found within **microtubule organizing centers (MTOCs)**. Centrioles are most active during cellular division. When a cell is ready to divide, the centrioles produce microtubules, which pull the replicated chromosomes apart and move them to opposite ends of the cell. Although centrioles are common in animal cells, they are not found in most types of plant cells.

Vacuoles

In Latin, the term *vacuole* means "empty cavity." But **vacuoles** are far from empty. They are fluid-filled sacs that store water, food, wastes, salts, or pigments. Vacuoles serve multiple functions in plant cells.

Peroxisomes

Peroxisomes are organelles that detoxify various substances, producing hydrogen peroxide (H_2O_2) as a byproduct. They also contain enzymes that break down hydrogen peroxide into oxygen and water. In animals, they are common in liver and kidney cells.

Cytoskeleton

Have you ever wondered what actually holds the cell together and enables it to keep its shape? The shape of a cell is determined by a network of protein fibers called the **cytoskeleton**. The most important fibers you'll need to know are **microtubules** and **microfilaments**.

Microtubules, which are made up of the protein **tubulin**, participate in cellular division and movement. These small fibers are an integral part of three structures: centrioles, **cilia,** and flagella. We've already mentioned that centrioles help chromosomes separate during cell division.

Microfilaments, like microtubules, are important for movement. These thin, rod-like structures are composed of the protein actin. Actin monomers are joined together and broken apart as needed to allow microfilaments to grow and shrink. Microfilaments assist during cytokinesis, muscle contraction, and formation of pseudopodia extensions during cell movement.

Cilia and Flagella

Cilia and flagella are threadlike structures best known for their locomotive properties in single-celled organisms. The beating motion of cilia and flagella structures propels these organisms through their watery environments.

The two classic examples of organisms with these structures are the *Euglena*, which gets about using its whiplike flagellum, and the *Paramecium*, which is covered in cilia. The rhythmic beating of the *Paramecium*'s cilia enables it to motor about in waterways, ponds, and microscope slides in your biology lab. You may have seen these in lab, but here's what they look like:

Euglena with one flagellum

Paramecium with many cilia

Though we usually associate such structures with microscopic or unicellular organisms, they aren't the only ones with cilia and flagella. As you probably know, these structures are also found in certain human cells. For example, the cells lining your respiratory tract possess cilia that sweep constantly back and forth (beating up to 20 times per second), helping to keep dust and unwanted debris from descending into your lungs. And every sperm cell has a type of flagellum that enables it to swim through the female reproductive organs to fertilize the waiting ovum.

Plant Cells Versus Animal Cells

Plant cells contain most of the same organelles and structures seen in animal cells, with several key exceptions. Plant cells, unlike animal cells, have a protective outer covering called the **cell wall** (made of cellulose). A cell wall is a rigid layer just outside the plasma membrane that provides support for the cell. It is found in plants, protists, fungi, and bacteria. Cell walls are important for protection against osmotic changes as well. In fungi, the cell wall is usually made of **chitin**, a modified polysaccharide. Chitin is also a principal component of an arthropod's exoskeleton. In addition, plant cells possess **chloroplasts**, which have a double outer membrane. Chloroplasts contain chlorophyll, the light-capturing pigment that gives plants and photosynthetic algae their characteristic green color. Because chloroplasts are involved in photosynthesis, we will discuss them in more detail in Chapter 6.

Did You Know?
Chloroplasts are found in plants and photosynthetic algae.

Another difference between plant and animal cells is that most of the cytoplasm within a plant cell is usually taken up by a large vacuole—the **central vacuole**—that crowds the other organelles. In mature plants, this vacuole contains the **cell sap**. A full vacuole in a plant is a sign that it is not dehydrated and has a healthy amount of turgor pressure. Dehydrated plants cannot fill their vacuoles and they wilt. Plant cells also differ from animal cells in that plant cells do not contain centrioles.

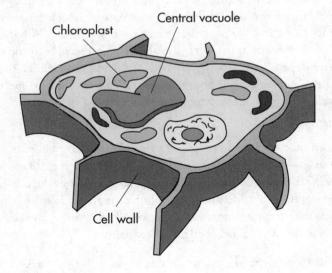

AP Biology Premium Prep

Make sure you learn these differences, as the AP Biology Exam almost always has a question about cellular structures.

Why do we need to know about the structure of cells? Because biological structure is often closely related to function. To help you remember the differences among prokaryotes, plant cells, and animal cells, we've put together a simple table.

Structural Characteristics of Different Cell Types			
Structure	**Prokaryote**	**Plant Cell**	**Animal Cell**
Cell Wall	Yes	Yes	No
Plasma Membrane	Yes	Yes	Yes
Membrane-Bound Organelles	No	Yes	Yes
Nucleus	No	Yes	Yes
Centrioles	No	No	Yes
Ribosomes	Yes	Yes	Yes

TRANSPORT: TRAFFIC ACROSS MEMBRANES

Homeostasis
Moving things across membranes is important for growth and homeostasis. Osmoregulation is the movement of water and solute across membranes to maintain homeostasis.

We've talked about the structure of cell membranes; now let's discuss how molecules and fluids pass through the plasma membrane, which separates the internal environment of the cell from the external environment. What are some of the patterns of membrane transport? The ability of molecules to move across the cell membrane depends on two things: (1) the semipermeability of the plasma membrane and (2) the size and charge of particles that want to get through. In plant cells, cell walls provide a structural permeability barrier made of complex carbohydrates.

Since the plasma membrane is composed primarily of phospholipids, small, lipid-soluble substances—such as N_2, O_2, and CO_2—cross the membrane without any resistance. Why? Because "like dissolves like." The lipid bilayer is a hydrophobic sandwich (hydrophilic outside and hydrophobic interior), and only hydrophobic things can pass that central zone. If a substance is hydrophilic, the bilayer won't let it pass without assistance, called **facilitated transport**.

Facilitated transport depends upon a number of proteins that act as tunnels through the membrane. Channels are very specialized types of tunnels that let only certain things through.

The most famous of these are **aquaporins**, which are water-specific channels. Although water is polar, there are typically sufficient aquaporins for water to traverse the membrane whenever it wishes without forming traffic jams. However, without aquaporins, no water would be able to cross the membrane. Glucose and ions such as Na^+ and K^+ are also transported across the plasma membrane via membrane proteins. Membranes may become polarized as these ions move across them.

Passive Transport: Simple and Facilitated Diffusion

Now we know how things cross the membrane, but let's talk about why they cross the membrane. The simple answer is "To get to the other side." If there is a high concentration of something in one area, it will move to spread out and diffuse into an area with a lower concentration, even if that means entering or exiting a cell. In other words, the substance moves *down* a concentration gradient. This is called diffusion. When the molecule that is diffusing is hydrophobic, the diffusion is called **simple diffusion** because the small nonpolar molecule can just drift right through the membrane without trouble. When the diffusion requires the help of a channel-type protein, it is called **facilitated diffusion**. Anytime that a substance is moving by diffusion, it is called **passive transport** because there is no outside energy required to power the movement. It's like riding a bicycle downhill. The molecules are just "going with the flow."

Passive transport does not require the input of metabolic energy. It is important because it helps the cell import resources and export wastes.

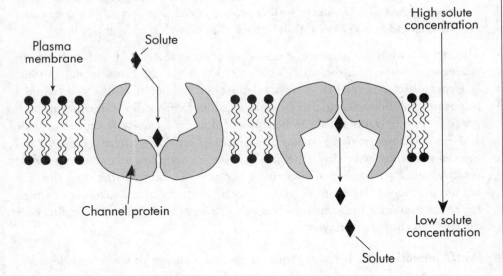

Osmosis

When the water is diffusing, the process is called **osmosis**. Water always wants to move from an area where it is most concentrated to an area where it is least concentrated. Water is not usually pure water; since it is so great at dissolving things, it usually exists as a solution. A solution is made when a liquid solvent dissolves solute particles. So, if we rethink water moving down its concentration gradient, this is the same as saying that it wants to go from where there is less solute (i.e., watery solution) to where there is more solute (i.e., concentrated solution). It's as if the water is moving to dilute the concentrated solute particles.

When discussing osmosis, the situation is often described such that there are two areas and the water can flow freely, but the solute cannot. For example, if a chamber containing water and a chamber containing a sucrose solution are connected by a semipermeable membrane that allows water but not sucrose to cross, diffusion of sucrose between the chambers cannot occur. In this case, osmosis draws water into the sucrose chamber to reduce the sucrose concentration. This will reduce the total volume of the water chamber. Water will flow into the sucrose chamber until the concentration is the same across the membrane.

In both diffusion and osmosis, the final result is that solute concentrations are the same on both sides of the membrane. The only difference is that in diffusion the membrane is usually permeable to solute, and in osmosis it is not.

In plants, the cell wall is important to protect it against osmotic changes. The cell wall stays relatively the same size all the time. On the other hand, the cell membrane can shrink away from the wall (a process called plasmolysis) if it loses water and can expand and squeeze tightly against the cell wall if it takes in water.

The term **tonicity** is used to describe osmotic gradients. If an environment is **isotonic** to the cell, the solute concentration is the same inside and outside. A **hypertonic** solution has more total dissolved solutes than the cell, while a **hypotonic** solution has less. The tendency of water to move down its concentration gradient (into or out of cells) can be a powerful force; it can cause cells to explode and can overcome gravity. You may also hear the terms *isosmotic*, *hyperosmotic*, and *hypoosmotic*. These words are generally used when comparing two solutions, while tonicity terms are used when comparing a solution to a cell. Other than this, the definitions are similar; isotonic and isosmotic both describe situations in which the concentration is the same on either side of a membrane, for example. The only difference is the type of membrane.

Water potential (Ψ) is the measure of potential energy in water and describes the eagerness of water to flow from an area of high water potential to an area of low water potential. It is affected by two factors: pressure potential (Ψ_p) and solute potential (Ψ_s). The equation for **solute (osmotic) potential** is listed on the AP Biology Equations and Formulas sheet and describes the effect of solute concentration on water flow.

Don't Forget the Formula Sheet!
Be sure to look at the Equations and Formulas sheet (pages 333–334) to be prepared for questions about water potential and solute potential.

Water Potential Equations

$$\Psi = \Psi_p + \Psi_s$$

Ψ_p = pressure potential

Ψ_s = solute potential

Solute Potential of a Solution

$$\Psi_s = -iCRT$$

where:

i = ionization constant

C = molar concentration

R = pressure constant

$$\left(R = 0.0831 \frac{L \cdot bars}{mol \cdot K} \right)$$

T = temperature in Kelvin (°C + 273)

Adding a **solute** lowers the water potential of a solution, causing water to be less likely to leave this solution and more likely to flow into it. This makes sense because the added solute makes the solution more concentrated, and the water is now unlikely to diffuse away. The more solute molecules present, the more negative the solute potential is.

A red blood cell dropped into a hypotonic solution (such as distilled water) will expand because water will move into the cell, an area of lower water potential. Eventually, the red blood cell will pop. If a similar experiment is done with a plant cell, water will still move into the cell, but the cell wall will exert pressure, increasing the water potential and limiting the gain of water. Water potential also explains how water moves from soil into plant roots and how plants transport water from roots to leaves to support photosynthesis.

> **Remember It Like This**
> An area of high water potential is an area of high water concentration (e.g., low solute).

Active Transport

Suppose a substance wants to move in the opposite direction—from a region of lower concentration to a region of higher concentration. A transport protein can help usher the substance across the plasma membrane, but it's going to need energy to accomplish this. This time it's like riding a bicycle uphill. Compared with riding downhill, riding uphill takes a lot more work. Movement against the natural flow is called **active transport**.

But where does the protein get this energy? Some proteins in the plasma membrane are powered by ATP. The best example of active transport is a special protein called the **sodium-potassium pump**. It ushers out three sodium ions (Na^+) and brings in two potassium ions (K^+) across the cell membrane. This pump depends on ATP to get ions across that would otherwise remain in regions of higher concentration. Primary active transport occurs when ATP is directly utilized to transport something. Secondary active transport occurs when something is actively transported using the energy captured from the movement of another substance flowing down its concentration gradient.

> **Resting Membrane Potential**
> The sodium-potassium pump is important for setting up the typical charge within the cell, and the inside is always a little bit negative. This resting membrane potential (difference in charge between inside and outside the cell) is important for passing signals in the body.

We've now seen that small substances can cross the cell membrane by:

- simple diffusion
- facilitated transport
- active transport

Endocytosis

When the particles that want to enter a cell are just too large, the cell uses a portion of the cell membrane to engulf the substance. The cell membrane forms a pocket, pinches in, and eventually forms either a vacuole or a vesicle. This process is called **endocytosis**.

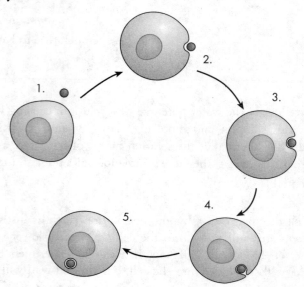

Three types of endocytosis exist: **pinocytosis**, **phagocytosis**, and **receptor-mediated endocytosis**. In pinocytosis, the cell ingests liquids ("cell-drinking"). In phagocytosis, the cell takes in solids ("cell-eating"). A special type of endocytosis, receptor-mediated endocytosis, involves cell surface receptors that work in tandem with endocytic pits that are lined with a protein called **clathrin**. When a particle, or ligand, binds to one of these receptors, the ligand is brought into the cell by the invagination, or "folding in" of the cell membrane. A vesicle then forms around the incoming ligand and carries it into the cell's interior.

Bulk Flow

Other substances move by **bulk flow**. Bulk flow is the one-way movement of fluids brought about by *pressure*. For instance, the movement of blood through a blood vessel and the movement of fluids in the xylem and phloem of plants are examples of bulk flow.

Dialysis

Dialysis is the diffusion of *solutes* across a selectively permeable membrane. Special membranes that have holes of a certain size within them can be used to sort substances by using the process of diffusion. Kidney dialysis is a specialized process by which the blood is filtered by using machines and concentration gradients. Things present at high levels will naturally diffuse out of the blood; dialysis just gives them the opportunity.

Exocytosis

Sometimes large particles are transported *out* of the cell. In **exocytosis**, a cell ejects waste products or specific secretion products, such as hormones, by the fusion of a vesicle with the plasma membrane, which then expels the contents into the extracellular space. Think of exocytosis as reverse endocytosis.

CHAPTER 5 KEY TERMS

cells
surface area-to-volume ratio
light microscopes
electron microscopes
prokaryotic cells
eukaryotic cells
cytoplasm
nucleoid
cell wall
plasma membrane
flagella
capsule
organelles
phospholipid bilayer
peripheral proteins
integral proteins
transmembrane proteins
fluid-mosaic model
adhesion proteins
receptor proteins
transport proteins
channel proteins
cell surface markers
carbohydrate side chains
nucleus
chromosomes
nucleolus
ribosomes
endoplasmic reticulum (ER)
rough ER
smooth ER
Golgi bodies
vesicles
mitochondria
adenosine triphosphate (ATP)
cristae
lysosomes
centrioles

microtubule organizing centers
   (MTOCs)
vacuoles
peroxisomes
cytoskeleton
microtubules
microfilaments
tubulin
cilia
Euglena
Paramecium
chitin
chloroplasts
central vacuole
cell sap
facilitated transport
aquaporins
simple diffusion
facilitated diffusion
passive transport
osmosis
tonicity
isotonic
hypertonic
hypotonic
water potential
solute (osmotic) potential
solute
active transport
sodium-potassium pump
endocytosis
pinocytosis
phagocytosis
receptor-mediated endocytosis
clathrin
bulk flow
dialysis
exocytosis

Chapter 5 Drill

Answers and explanations can be found in Chapter 15.

1   ☐ Mark for Review

A researcher collects a sample of water from a pond that contains numerous microscopic cells containing green pigments that are covered in a mucilaginous sheath. The cells contain considerable amounts of chlorophyll *a* and phycobilin pigments, but they do not contain organized nuclei. If the researcher analyzes the cells using an electron microscope, which other organelles will also be present?

- Ⓐ A cell wall and DNA lacking histones
- Ⓑ Lysosomes and mitochondria
- Ⓒ Golgi bodies and chloroplasts
- Ⓓ Ribosomes and glyoxysomes

2   ☐ Mark for Review

The development of electron microscopy has provided key insights into many aspects of cellular structure and function, which had previously been too small to be seen. Which of the following would require the use of electron microscopy for visualization?

- Ⓐ The cells making up the stem of a flower
- Ⓑ The cells swabbed from an ear to treat an ear infection
- Ⓒ The pores on the nuclear membrane
- Ⓓ The shape and arrangement of bacterial cells

3   ☐ Mark for Review

The antibiotic cephalosporin is lethal to certain bacteria cells, as it prevents the cells from forming cell walls and ultimately causes the cells to burst. Based on this information, what type of medium do the bacteria cells live in?

- Ⓐ Isotonic
- Ⓑ Hypotonic
- Ⓒ Hypertonic
- Ⓓ One with an identical osmotic concentration to the cell

5. Cell Structure and Function | **135**

Questions 4 through 5 refer to the following.

Ten equal-sized pieces of skinned eggplant are placed in solutions of different sodium chloride concentrations and left to soak overnight. The eggplant pieces are massed immediately before and after soaking, and the results are graphed in the figure below.

4 Mark for Review

Which of the following occurs in the cells of the eggplant sample that is placed in a 0.1 M concentration of sodium chloride solution?

(A) The cells of the sample shrink.

(B) The cells of the sample become flaccid.

(C) The volume of cytoplasm inside the cells decreases.

(D) The volume of the contents inside the vacuoles of the cells increases.

5 Mark for Review

What is the approximate concentration of the eggplant samples?

(A) 0.0 M

(B) 0.3 M

(C) 0.6 M

(D) 0.9 M

6 ☐ Mark for Review

In order to treat stomach flu due to rotavirus, medical practitioners typically prescribe electrolytes to replace fluids lost to nausea and diarrhea. Patients are sometimes given an antibiotic, tetracycline, to treat the symptoms. Tetracycline obstructs the ribosomes of the bacteria that typically cause the stomach flu. Which of the following explains why the antibiotic does not negatively impact humans?

(A) The antibiotic is not able to pass through eukaryotic cell membranes into human cells.

(B) The antibiotic affects eukaryotic ribosomes as well as prokaryotic ribosomes, but this doesn't have any effect on eukaryotic cells.

(C) Human cells are eukaryotic, which do not contain any ribosomes for the antibiotic to target.

(D) Bacteria are prokaryotes, which have different ribosomes than eukaryotes that are targeted by tetracycline.

7 ☐ Mark for Review

A spherical cell has a volume of 113.04 cm³ and a surface area of 112.89 cm³. A cylindrical cell also has a volume of 113.04 cm³ and has a surface area of 755.48 cm³. Which cell will be able to maintain its temperature more rapidly?

(A) The spherical cell because it has a higher surface area-to-volume ratio.

(B) The spherical cell because it has a lower surface area-to-volume ratio.

(C) The cylindrical cell because it has a higher surface area-to-volume ratio.

(D) The cylindrical cell because it has a lower surface area-to-volume ratio.

8 ☐ Mark for Review

Cells come in a variety of shapes and sizes. Lymphocytes are spherical, red blood cells are disc-shaped, and muscle and nerve cells are cylindrical. Which of the following correctly matches a cell type with an advantage of its shape?

(A) Lymphocytes are spherical, which makes it easier for them to move through blood vessels to target pathogens.

(B) Red blood cells are rod-shaped, which makes them flexible and allows them to take in oxygen and carbon dioxide.

(C) Nerve cells are cylindrical, which makes it easier for them to travel through spinal fluid to repair injuries.

(D) Muscle cells are spherical, which makes it easier for cells to send signals to other muscles.

9 ☐ Mark for Review

Which of the following would have the greatest concentration of rough endoplasmic reticulum present inside of it?

(A) A bioluminescent bacteria species

(B) A liver cell involved in the creation of enzymes

(C) A xylem cell transporting neurotransmitters

(D) An amoeba consuming small particles

Chapter 5 Summary

- Living things are composed of cells, the smallest unit of living organisms. Cells may be categorized into prokaryotes, which do not have a nucleus or membrane-bound organelles, and eukaryotes, which have a nucleus and membrane-bound organelles. Features of prokaryotic cells (like bacteria, for example) can include the following:
    - plasma membrane
    - flagella and cilia
    - cell wall
    - nucleoid region
    - ribosomes
    - circular chromosome
    - cytoplasm

- Components of the eukaryotic cells include the following:
    - nucleus
    - nucleolus
    - rough endoplasmic reticulum
    - smooth endoplasmic reticulum
    - Golgi body
    - ribosomes
    - mitochondria
    - vacuole
    - lysosome
    - peroxisomes
    - centrioles (animals only)
    - cytoplasm
    - plasma membrane
    - cytoskeleton (microtubules, microfilaments)
    - central vacuole (plants only)
    - cell wall (plants, fungi, and some protists)

- The plasma membrane is composed of a phospholipid bilayer. The hydrophobic nature of the inside of the membrane is responsible for its selective permeability. Transport through the membrane is dependent on size and polarity of the molecules and concentration gradients. There are a few different types of cellular transport:
    - simple diffusion
    - facilitated transport
    - active transport
    - bulk flow
    - dialysis
    - endocytosis (pinocytosis, phagocytosis, receptor-mediated endocytosis)
    - exocytosis

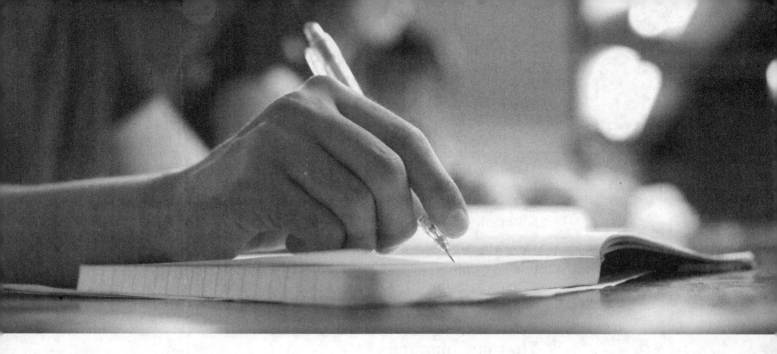

Chapter 6
Cellular Energetics

BIOENERGETICS

In Chapter 4, we discussed some of the more important organic molecules. But what makes these molecules so important? Glucose, starch, and fat are all energy-rich. However, the energy is packed in chemical bonds holding molecules together. To carry out the processes necessary for life, cells must find a way to release the energy in these bonds when they need it and store it away when they don't. The study of how cells accomplish this is called **bioenergetics**. Generally, bioenergetics is the study of how energy from the sun is transformed into energy in living things.

During chemical reactions, such bonds are either broken or formed. This process involves energy, no matter which direction we go. Every chemical reaction involves a change in energy.

Thermodynamics

We've all heard the expression *"nothing in life is free."* The same holds true in nature.

> Energy cannot be created or destroyed. In other words, the sum of energy in the universe is constant.

This rule is called the **First Law of Thermodynamics**. As a result, the cell cannot take energy out of thin air. Rather, it must harvest it from somewhere.

The **Second Law of Thermodynamics** states that energy transfer leads to less organization. That means the universe tends toward disorder (**entropy**).

In order to power cellular processes, energy input must exceed energy loss to maintain order. Cellular processes that release energy can be coupled with cellular processes that require an input of energy. Many pathways are sequential to make energy transfer more efficient. A product of one step is often a reactant for the next step in a process. A lack of energy flow or order results in death for an organism.

Types of Reactions

Exergonic reactions are those in which the products have *less* energy than the reactants. Simply put, energy is given off during the reaction.

Let's look at an example. The course of a reaction can be represented by an **energy diagram**. Here's an energy diagram for an exergonic reaction.

You'll notice that energy is represented along the *y*-axis. Based on the diagram, we see that our reaction released energy. An example of an exergonic reaction is the oxidizing of molecules in mitochondria of cells, which then releases the energy stored in the chemical bonds.

Reactions that require an input of energy are called **endergonic reactions**. You'll notice that the products have *more* energy than the reactants. An example is plants' use of carbon dioxide and water to form sugars.

Activation Energy

Even though exergonic reactions release energy, the reaction might not occur naturally without a little bit of energy to get things going. This is because the reactants must first turn into a high energy molecule, called the **transition state**, before turning into the products. The transition state is sort of a reactants-products hybrid state that is difficult to achieve. In order to reach this transition state, a certain amount of energy is required. This is called the **activation energy**.

Be Careful
A transition state is not the same thing as a reaction intermediate. Transition states occur in each step of every reaction, between the reactant(s) and the product(s). Intermediates are formed between each step of a multistep reaction.

You'll notice that we needed a little energy to get us going. That's because chemical bonds must be broken before new bonds can form. This energy barrier—the hump in the graph—is called the activation energy. Once a set of reactants has reached its activation energy, the rest of the reaction is all downhill. Reaching the transition state is the tough part.

ENZYMES

A catalyst is something that speeds something up. **Enzymes** are biological catalysts that speed up reactions. They accomplish this by lowering the activation energy and helping the transition state to form. Thus, the reaction can occur more quickly because the tricky transition state is not as much of a hurdle to overcome. Enzymes do NOT change the energy of the starting point or the ending point of the reaction. They only lower the activation energy.

Enzyme Specificity

Most of the crucial reactions that occur in the cell require enzymes. Yet enzymes themselves are highly specific—in fact, each enzyme catalyzes only one kind of reaction. This is known as **enzyme specificity**. Because of this, enzymes are usually named after the molecules they target. In enzymatic reactions, the targeted molecules are known as **substrates**. For example, maltose, a disaccharide, can be broken down into two glucose molecules. Our substrate, maltose, gives its name to the enzyme that catalyzes this reaction: maltase.

Many enzymes are named simply by replacing the suffix of the substrate with *-ase*. Using this nomenclature, malt*ose* becomes malt*ase*.

Enzyme-Substrate Complex

Enzymes have a unique way of helping reactions along. As we just saw, the reactants in an enzyme-assisted reaction are known as substrates. During a reaction, the enzyme's job is to bring the transition state about by helping the substrate(s) get into position. It accomplishes this through a special region on the enzyme known as an **active site**.

The enzyme temporarily binds one or more of the substrates to its active site and forms an **enzyme-substrate complex**. In general, the structure of the substrate—specifically its shape and charge—must be compatible with the active site of the enzyme in order for the enzyme-catalyzed reaction to occur. Let's take a look:

Once the reaction has occurred and the product is formed, the enzyme is released from the complex and restored to its original state. Now the enzyme is free to react again with another bunch of substrates.

By binding and releasing substrates over and over again, the enzyme speeds the reaction along, enabling the cell to release much-needed energy from various molecules. Here is a quick review on the function of enzymes.

Enzymes Do:
- increase the rate of a reaction by lowering the reaction's activation energy
- form temporary enzyme-substrate complexes
- remain unaffected by the reaction

Enzymes Don't:
- change the reaction
- make reactions occur that would otherwise not occur at all

Induced-Fit

However, scientists have discovered that enzymes and substrates don't fit together quite so seamlessly. It appears that the enzyme has to change its shape slightly to accommodate the shape of the substrates. This is called **induced-fit**. Sometimes certain factors are involved in activating the enzyme and making it capable of binding the substrate. Because the fit between the enzyme and the substrate must be perfect, enzymes operate only under a strict set of biological conditions.

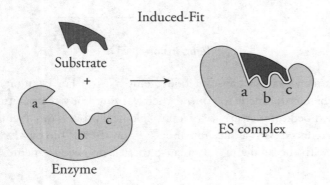

Enzymes Don't Always Work Alone

Enzymes sometimes need a little help in catalyzing a reaction. Those factors are known as **cofactors**. Cofactors can be either organic molecules (called coenzymes), inorganic molecules, or ions. Inorganic cofactors are usually metal ions (Fe^{2+}, Mg^{2+}). Vitamins are examples of organic coenzymes. Your daily dose of vitamins is important for just this reason: vitamins are active and necessary participants in crucial chemical reactions.

Factors Affecting Reaction Rates

Enzymatic reactions can be influenced by a number of factors, such as temperature and pH. The concentrations of enzyme and substrate will also determine the speed of the reaction.

Temperature

The rate of a reaction increases with increasing temperature, up to a point, because an increase in the temperature of a reaction increases the frequency of collisions among the molecules. But too much heat can damage an enzyme. If a reaction is conducted at an excessively high temperature (above 42°C), the enzyme loses its three-dimensional shape and becomes inactivated. Enzymes damaged by heat and deprived of their ability to catalyze reactions are said to be **denatured**, or disrupted. In some cases, however, enzyme denaturation is reversible if the original optimal environmental conditions of the enzyme are restored. This allows the enzyme to regain its activity and function.

Here's one thing to remember: all enzymes operate at an ideal temperature. For most human enzymes, this temperature is body temperature, 37°C.

Break Down Words
To denature is to "un-nature" something, to change its nature, or to remove the characteristics that make it unique. Without proper and special folding, a protein loses its nature.

Not all organisms have a constant body temperature. For example, ectotherms depend on the environment to control their varying body temperatures. Q_{10} is a measure of temperature sensitivity of a physiological process or enzymatic reaction rate. If you are faced with a question about Q_{10} on the AP Biology Exam, you will be provided with the following formula, so there is no need to memorize it.

$$Q_{10} = \left(\frac{k_2}{k_1}\right)^{\frac{10}{T_2 - T_1}}$$

The temperature unit must be in either Celsius or Kelvin, and the same unit must be used for T_1 and T_2. The two reaction rates (k_1 and k_2) must also have the same unit. Q_{10} has no unit; it is the factor by which the rate of a reaction increases due to a temperature increase. The more temperature-dependent a reaction is, the higher Q_{10} will be. Reactions with $Q_{10} = 1$ are temperature independent.

pH
Enzymes also function best at a particular pH. At an incorrect pH, the hydrogen bonds can be disrupted and the structure of the enzyme can be altered. For most enzymes, the optimal pH is at or near a pH of 7.

Other enzymes operate at a low pH. For instance, pepsin, the digestive enzyme found in the stomach, is most effective at an extremely acidic pH of 2. The enzymes in the lysosome also function best in an acidic pH.

Relative Concentration of Substrates and Products
The relative concentration of substrates and products can also affect the rate of an enzyme-catalyzed reaction. For example, an increase in substrate concentration will initially speed up the reaction. However, once all of the enzyme in solution is bound by substrate, the reaction can no longer speed up. This concentration of substrate where all of the enzyme in a reaction is bound by substrate is called the **saturation point**. Additional substrate past this point will no longer increase the speed of the reaction.

Enzyme Regulation
We know that enzymes control the rates of chemical reactions. But what regulates the activity of enzymes? It turns out that a cell can control enzymatic activity by regulating the conditions that influence the shape of the enzyme. Enzymes can be turned on or off by things that bind to them. Sometimes these things can bind at the active site, and sometimes they bind at other sites, called **allosteric sites**.

If the substance has a shape that fits the active site of an enzyme (i.e., similar to the substrate or the transition state), it can compete with the substrate and block the substrate from getting into the active site. This is called **competitive inhibition**. If there was enough substrate available, the substrate would out-compete the inhibitor and the reaction would occur. You can always identify a competitive inhibitor based on what happens when you flood the system with lots of substrate.

If the inhibitor binds to an allosteric site, it is an **allosteric inhibitor** and it is **noncompetitive inhibition**. A noncompetitive inhibitor generally distorts the enzyme shape so that it cannot function. The substrate can still bind at the active site, but the enzyme will not be able to catalyze the reaction.

REACTION COUPLING AND ATP

As we just saw, almost everything an organism does requires energy. How, then, can the cell acquire the energy it needs without becoming a major mess? Fortunately, it's through adenosine triphosphate (ATP).

ATP, as the name indicates, consists of a molecule of adenosine bonded to three phosphates. The great thing about ATP is that an enormous amount of energy is packed into those phosphate bonds.

Not all molecules that bind to the allosteric site of an enzyme are inhibitors. Some enzymes are activated by allosteric regulators.

The body can perform difficult endergonic reactions by coupling them to ATP hydrolysis, which is very exergonic.

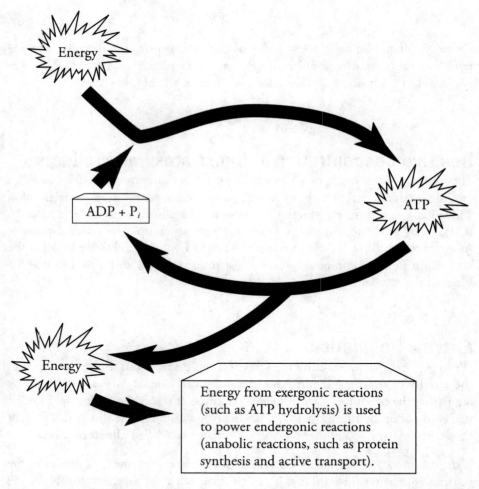

When a cell needs energy, it takes one of these potential-packed molecules of ATP and splits off the third phosphate, forming adenosine diphosphate (ADP) and one loose phosphate (P_i), while releasing energy in the process.

$$ATP \rightarrow ADP + P_i + energy$$

The energy released from this reaction can then be put to whatever use the cell pleases. Of course, this doesn't mean that the cell is above the laws of thermodynamics. But within those constraints, ATP is the best source of energy the cell has available. It is relatively neat (only one bond needs to be broken to release that energy) and relatively easy to form. Organisms can use exergonic processes that increase energy, like breaking down ATP, to power endergonic reactions, like building organic macromolecules.

Sources of ATP

But where does all this ATP come from? It can be formed in several ways, but the bulk of it comes from a process called **cellular respiration**. Cellular respiration is a process of breaking down sugar and making ATP.

In autotrophs, the sugar is made during **photosynthesis**. In heterotrophs, glucose comes from the food we eat. In plants, photosynthesis is a precursor to cellular respiration. So let's start there.

PHOTOSYNTHESIS

Plants, algae, and cyanobacteria are producers. The earliest photosynthesis was conducted by prokaryotic cyanobacteria, and later, eukaryotes developed photosynthetic abilities. All they do is bask in the Sun, churning out the glucose necessary for life. In this section, we'll focus on how plants conduct photosynthesis.

As we discussed earlier, photosynthesis is the process by which light energy is converted to chemical energy. Here's an overview of photosynthesis.

$$6CO_2 + 6H_2O \longrightarrow C_6H_{12}O_6 + 6O_2$$

You'll notice from this equation that carbon dioxide and water are the raw materials used to manufacture simple sugars. You'll also see that oxygen is one of the products of photosynthesis. One important concept to know for the AP Exam is that there is strong evidence that prokaryotic photosynthesis contributed to the production of oxygen in the atmosphere. Furthermore, prokaryotic photosynthesis pathways laid the evolutionary foundation for eukaryotic photosynthesis to develop. But remember, there's much more to photosynthesis than what's shown in the simple reaction above.

There are two stages in photosynthesis: the **light reactions** (also called the light-dependent reactions) and the **dark reactions** (also called the light-independent reactions). The whole process begins when **photons** (energy units) of sunlight strike a leaf, activating chlorophyll and exciting electrons. The activated chlorophyll molecule then passes these excited electrons down to a series of electron carriers, ultimately producing ATP and NADPH. Both of these products, along with carbon dioxide, are then used in the dark reactions (light-independent) to make carbohydrates. Along the way, water is also split and oxygen gets released.

Algae are a diverse group of aquatic, eukaryotic organisms that can perform photosynthesis. Most algae are protists. Some common examples are seaweed (such as kelp or phytoplankton), pond scum, or algal blooms in lakes.

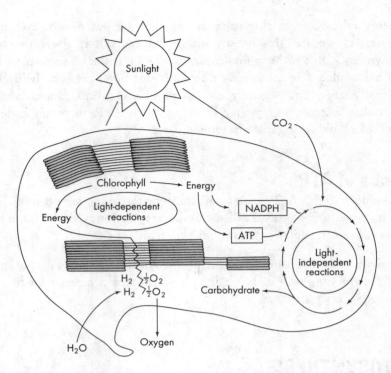

We'll soon see that this beautifully orchestrated process occurs thanks to a whole host of special enzymes and pigments. But before we turn to the stages in photosynthesis, let's talk about where photosynthesis occurs.

Chloroplast Structure

The leaves of plants contain lots of chloroplasts, which are the primary sites of photosynthesis.

Now let's look at an individual chloroplast. If you split the membranes of a chloroplast, you'll find a fluid-filled region called the **stroma**. Inside the stroma are structures that look like stacks of coins. These structures are the **grana**.

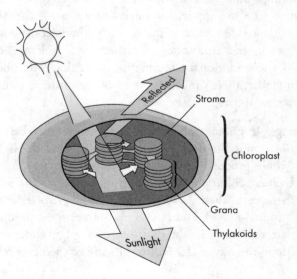

The many disk-like structures that make up grana are called **thylakoids.** They contain chlorophyll, a light-absorbing pigment that drives photosynthesis, as well as enzymes involved in the process. The very inside of a thylakoid is called the thylakoid lumen.

Many light-absorbing pigments participate in photosynthesis. Some of the more important ones are **chlorophyll *a***, **chlorophyll *b***, and **carotenoids**. These pigments are clustered in the thylakoid membrane into units called antenna complexes.

All of the pigments within a unit are able to "gather" light, but they're not able to "excite" electrons. Only one special molecule—located in the **reaction center**—is capable of transforming light energy to chemical energy. In other words, the other pigments, called **antenna pigments**, "gather" light and "bounce" energy to the reaction center.

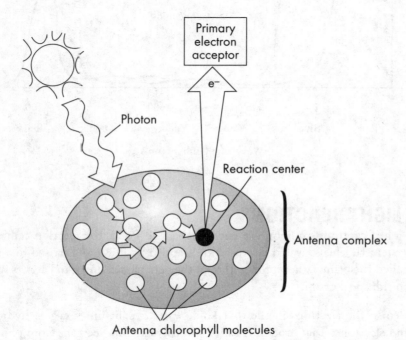

There are two types of reaction centers: **photosystem I (PS I)** and **photosystem II (PS II)**. The principal difference between the two is that each reaction center has a specific type of chlorophyll—chlorophyll *a*—that absorbs a particular wavelength of light. For example, **P680**, the reaction center of photosystem II, has a maximum absorption at a wavelength of 680 nanometers. The reaction center for photosystem I —**P700**—best absorbs light at a wavelength of 700 nanometers.

When light energy is used to make ATP, it is called **photophosphorylation**. Autotrophs are using light (that's *photo*) and ADP and phosphates (that's *phosphorylation*) to produce ATP.

An **absorption spectrum** shows how well a certain pigment absorbs electromagnetic radiation. Light absorbed is plotted as a function of radiation wavelength. This spectrum is the opposite of an **emission spectrum**, which gives information on which wavelengths are emitted by a pigment. The absorption spectrum for chlorophyll *a*, chlorophyll *b,* and carotenoids is below. You can see that chlorophyll *a* and chlorophyll *b* absorb blue and red light quite well but do not absorb light in the green part of the spectrum. Light in the green range of wavelengths is reflected, and this is why chlorophyll and many plants are green. Carotenoids absorb light on the blue-green end of the spectrum, but not on the other end. This is why plants rich in carotenoids are yellow, orange, or red.

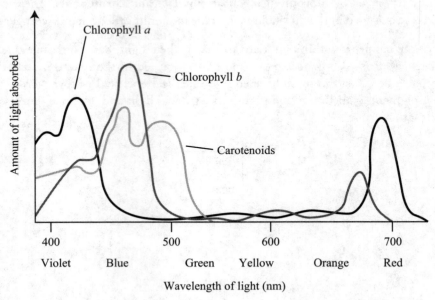

THE LIGHT REACTIONS

When a leaf captures sunlight, the energy is sent to P680, the reaction center for photosystem II. The activated electrons are trapped by P680 and passed to a molecule called the primary acceptor, and then they are passed down to carriers in the electron transport chain.

To replenish the electrons in the thylakoid, water is split into oxygen, hydrogen ions, and electrons. That process is called **photolysis**. The electrons from photolysis replace the missing electrons in photosystem II.

As the energized electrons from photosystem II travel down the electron transport chain, they pump hydrogen ions into the thylakoid lumen. A proton gradient is established. As the hydrogen ions move back into the stroma through ATP synthase, ATP is produced.

After the electrons leave photosystem II, they go to photosystem I. The electrons are passed through a second electron transport chain until they reach the final electron acceptor NADP⁺ to make **NADPH**.

> The light reactions of photosynthesis occur in the thylakoid membranes.

Here is a summary of the most important points from the first part of photosynthesis:

- P680 in photosystem II captures light and passes excited electrons down an electron transport chain to produce ATP.
- P700 in photosystem I captures light and passes excited electrons down an electron transport chain to produce NADPH.
- A molecule of water is split by sunlight, releasing electrons, hydrogen, and free O_2.
- The light-dependent reactions occur in the grana of chloroplasts, where the thylakoids are found. Remember: the light-absorbing pigments and enzymes for the light-dependent reactions are found within the thylakoids.

Most plants follow this electron flow, which is called linear, or noncyclic. However, some plants (such as C_4 plants discussed below) perform cyclic electron flow instead. Cyclic photophosphorylation is similar to noncyclic, but generates only ATP, and no NADPH. It takes place only at photosystem I: once an electron is displaced from the photosystem, it is passed down electron acceptor molecules and returns to photosystem I, from where it was emitted.

Photosystem I and photosystem II were numbered in order of their discovery, not the order they are used in photosynthesis.

The Light-Independent Reactions

Now let's turn to the dark side. The dark reactions use the products of the light reactions—ATP and NADPH—to make sugar. We now have energy to make glucose, but what do plants use as their carbon source? CO_2, of course. You've probably heard of the term **carbon fixation**. All this means is that CO_2 from the air is converted into carbohydrates. This step occurs in the stroma of the leaf. The dark reactions are also called the **Calvin-Benson Cycle** (or just Calvin cycle).

Let's summarize important facts about the dark reactions:

- The Calvin cycle occurs in the stroma of chloroplasts.
- ATP and NADPH from the light reactions are necessary for carbon fixation.
- CO_2 is fixed to form glucose.

Photosynthesis: A Summary

The following table summarizes the stages of photosynthesis:

Stage of Photosynthesis	Location	Input	Output
Light-Dependent Reactions	Thylakoid membrane of chloroplasts Photosystem II (P680) Photosystem I (P700)	Photons H_2O	NADPH ATP O_2
Light-Independent Reactions (Calvin cycle)	Stroma	$3CO_2$ 9 ATP 6 NADPH	sugar

The light-dependent and light-independent reactions are inexorably linked; neither set of reactions alone can produce carbohydrates from CO_2. The light-dependent reactions use water and light to produce ATP, NADPH, and O_2; the light-independent reactions use CO_2, ATP, and NADPH to produce ADP, $NADPH^+$, and carbohydrates.

Stomata are pores on the leaf surface that allow CO_2 to enter the leaf and O_2 and water to exit. Most plants close their stomata on hot, dry days to prevent water loss by transpiration (the evaporative loss of water from leaves). While this conserves water, it also limits access to CO_2 and thus reduces photosynthetic yield. With less CO_2 available, O_2 accumulates and plants start performing photorespiration instead of photosynthesis. **Photorespiration** is a wasteful process that uses ATP and O_2, produces more CO_2, and doesn't produce any sugars.

Plants that live in hot climates have evolved two different ways around this: **CAM plants** deal with this problem by temporally separating carbon fixation and the Calvin cycle. They open their stomata at night and incorporate CO_2 into organic acids. During the day, they close their stomata and release CO_2 from the organic acids while the light reactions run. In contrast, **C_4 plants** have slightly different leaf anatomy that allows them to perform CO_2 fixation in a different part of the leaf from the rest of the Calvin cycle. This prevents photorespiration. C_4 plants produce a four-carbon molecule as the first product of carbon fixation and perform cyclic electron flow in the light reactions (discussed on the previous page).

CELLULAR RESPIRATION

In the shorthand version, cellular respiration looks something like this:

$$C_6H_{12}O_6 + 6O_2 \rightarrow 6CO_2 + 6H_2O + ATP$$

Notice that we've taken a sugar, perhaps a molecule of glucose, and combined it with oxygen to produce carbon dioxide, water, and energy in the form of our old friend, ATP. However, as you probably already know, the actual picture of what really happens is far more complicated.

Generally speaking, we can break cellular respiration down into two different approaches: aerobic respiration and anaerobic respiration. If ATP is made in the presence of oxygen, we call it **aerobic respiration**. If oxygen isn't present, we call it **anaerobic respiration**. Let's jump right in with aerobic respiration.

Introduction to Aerobic Respiration

Aerobic respiration consists of four stages containing a series of coupled reactions that establish an electrochemical gradient across membranes:

1. glycolysis
2. formation of acetyl-CoA
3. the Krebs (citric acid) cycle
4. oxidative phosphorylation (the electron transport chain + chemiosmosis)

In the first three stages, glucose is broken down and energy molecules are made. Some of these are ATP, and others are special electron carriers called **NADH** and **$FADH_2$**. In the fourth stage, these electron carriers unload their electrons, and the energy is eventually used to make even more ATP.

Stage 1: Glycolysis

The first stage begins with **glycolysis**, the splitting (*-lysis*) of glucose (*glyco-*). Glucose is a six-carbon molecule that is broken into two three-carbon molecules called **pyruvic acid**. This breakdown of glucose also results in the net production of two molecules of ATP and two molecules of NADH.

Pyruvic acid is an acid. In your cells, it usually loses a proton, so it is called pyruvate instead.

$$\text{Glucose} + 2\ \text{ATP} + 2NAD^+ \rightarrow 2\ \text{Pyruvic acid} + 4\ \text{ATP} + 2NADH$$

AP Biology Premium Prep

Although we've written glycolysis as if it were a single reaction, this process doesn't occur in one step. In fact, it requires a sequence of enzyme-catalyzed reactions.

Let's look at a summary of these reactions:

Glycolysis occurs in the cytoplasm and starts with one glucose and ends with two pyruvates. There is a net production of 2 ATP and 2 NADH.

Summary
2 ATP used (investment)
4 ATP gained
2 NADH gained

If you take a good look at the reactions above, you'll see two ATP are needed to produce four ATP. You've probably heard the expression "You have to spend money to make money." In biology, you have to invest ATP to make ATP: our investment of two ATP yielded four ATP, for a net gain of two. The ATP molecules generated in glycolysis are created by combining ADP and an inorganic phosphate with the help of an enzyme.

Glycolysis also creates two NADH, which result from the transfer of electrons to the carrier NAD⁺, which then becomes NADH. NAD⁺ and NADH are constantly being turned into each other as electrons are being carried and then unloaded.

There are four important tidbits to remember regarding glycolysis:

- occurs in the cytoplasm
- net of 2 ATPs produced
- 2 pyruvic acids formed
- 2 NADH produced

Electron Carriers

$NAD^+ + H^+ + 2$ electrons = NADH

$FAD + 2H^+ + 2$ electrons = $FADH_2$

To carry electrons is to carry energy.

154 | For more free content, visit PrincetonReview.com

Stage 2: Formation of Acetyl-CoA

Pyruvic acid is transported to the mitochondrion. Each pyruvic acid (a three-carbon molecule) is converted to **acetyl coenzyme A** (a two-carbon molecule, usually just called acetyl-CoA) and CO_2 is released.

2 Pyruvic acid + 2 Coenzyme A + 2NAD⁺ → 2 Acetyl-CoA + $2CO_2$ + 2NADH

Are you keeping track of our carbons? We've now gone from two three-carbon molecules to two two-carbon molecules. The extra carbons leave the cell in the form of CO_2. Once again, two molecules of NADH are also produced for each glucose you started with. This process of turning pyruvic acid into acetyl-CoA is catalyzed by an enzyme complex called the **pyruvate dehydrogenase complex** (**PDC**).

> In the mitochondria, pyruvate is turned into acetyl-CoA and 1 NADH is made; double this if you are counting per glucose.

Stage 3: The Krebs Cycle

The next stage is the **Krebs cycle**, also known as the **citric acid cycle**. Each of the two acetyl coenzyme A molecules will enter the Krebs cycle, one at a time, and all the carbons will ultimately be converted to CO_2, which is released as a byproduct. This stage occurs in the **matrix** of the mitochondria.

The Krebs cycle begins with each molecule of acetyl-CoA produced from the second stage of aerobic respiration combining with **oxaloacetate**, a four-carbon molecule, to form a six-carbon molecule, **citric acid** or citrate (hence its name, the citric acid cycle).

Citrate gets turned into several other things, and because the cycle begins with a four-carbon molecule, oxaloacetate, it eventually gets turned back into oxaloacetate to maintain the cycle by joining with the next acetyl-CoA coming down the pipeline.

> The Krebs cycle occurs in the mitochondrial matrix. It begins with acetyl-CoA joining with oxaloacetate to make citric acid and ends with oxaloacetate, 1 ATP, 3 NADH, and 1 $FADH_2$; double this if you are counting per glucose.

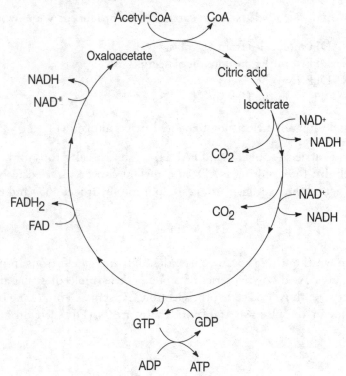

With each turn of the cycle, three types of energy are produced:

- 1 ATP
- 3 NADH
- 1 $FADH_2$

To figure out the total number of products per molecule of glucose, we simply double the number of products—after all, we started off the Krebs cycle with two molecules of acetyl-CoA for each molecule of glucose!

Now we're ready to tally up the number of ATP produced.

So far, we've made only four ATP—two ATP from glycolysis and two ATP from the Krebs cycle.

Although that seems like a lot of work for only four ATP, we have also produced hydrogen and electron carriers in the form of 10 NADH and 2 $FADH_2$. These molecules will, in turn, produce lots of ATP in the next stage of cellular respiration.

Stage 4: Oxidative Phosphorylation

Electron Transport Chain

As electrons (and the hydrogen atoms to which they belong) are removed from a molecule of glucose, they carry with them much of the energy that was originally stored in their chemical bonds. These electrons—and their accompanying energy—are then transferred to readied hydrogen carrier molecules. In the case of cellular respiration, these charged carriers are the coenzymes NADH and $FADH_2$.

> Electron transport chain reactions occur in chloroplasts, mitochondria, and prokaryotic plasma membranes. The electron transport chain here is similar to that seen in photosynthesis. They differ in their final electron acceptors: $NADP^+$ and O_2.

Let's see how many "loaded" electron carriers we've produced. We now have:

- 2 NADH molecules from glycolysis
- 2 NADH from the production of acetyl-CoA
- 6 NADH from the Krebs cycle
- 2 $FADH_2$ from the Krebs cycle

That gives us a total of 12 electron or energy carriers altogether.

These electron carriers—NADH and $FADH_2$—"shuttle" electrons to the **electron transport chain**; the resulting NAD^+ and FADH can be recycled to be used as carriers again, and the hydrogen atoms are split into hydrogen ions and electrons.

$$H_2 \rightarrow 2H^+ + 2e^-$$

Then, two interesting things occur. First, the high-energy electrons from NADH and $FADH_2$ are passed down a series of protein carrier molecules that are embedded in the cristae; thus, it is called the electron transport chain. Some of the carrier molecules in the electron transport chain are NADH dehydrogenase and **cytochrome C**.

Each carrier molecule hands down the electrons to the next molecule in the chain. The electrons travel down the electron transport chain until they reach the final electron acceptor, oxygen. Oxygen combines with these electrons (and some hydrogens) to form water. This explains the "aerobic" in aerobic respiration. If oxygen weren't available to accept the electrons, they wouldn't move down the chain at all, thereby shutting down the whole process of electron transport.

Chemiosmosis

At the same time that electrons are being passed down the electron transport chain, another mechanism is at work. Remember those hydrogen ions (also called *protons*) that split off from the original hydrogen atom? The energy released from the electron transport chain is used to pump hydrogen ions across the inner mitochondrial membrane from the matrix into the intermembrane space. The pumping of hydrogen ions into the intermembrane space creates a **pH gradient**, or **proton gradient**. The hydrogen ions really want to diffuse back into the matrix. The potential energy established in this gradient is responsible for the production of ATP. This pumping of ions and diffusion of ions to create ATP is **chemiosmosis**.

> Oxygen is required as the final stop in the electron transport chain.
>
> $NADP^+$ is the terminal electron acceptor in photosynthesis, while oxygen is the terminal electron acceptor in cellular respiration.
>
> Brown fat, which is found in newborns and hibernating mammals, can also use the hydrogen gradient generated from the electron transport chain to generate heat. This occurs when they decouple the chemiosmosis from oxidative phosphorylation.

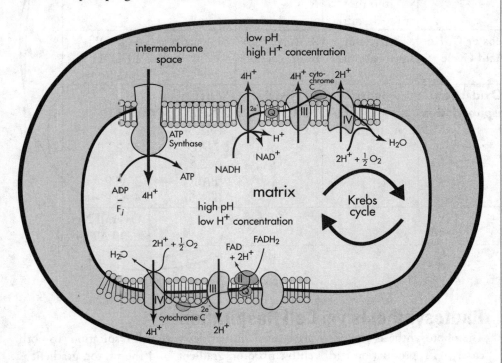

These hydrogen ions can diffuse across the inner membrane only by passing through channels called **ATP synthase**. Meanwhile, ADP and P_i are on the other side of these protein channels. The flow of protons through these channels produces ATP by combining ADP and P_i on the matrix side of the channel. Overall, this process is called **oxidative phosphorylation** because when electrons are given up, it is called "oxidation," and then ADP is "phosphorylated" to make ATP.

You're expected to know the following two things for the AP Biology Exam:

- Every NADH from glycolysis yields 1.5 ATP and all other NADH molecules yield 2.5 ATP.
- Every $FADH_2$ yields 1.5 ATP.

You will also want to make sure you remember the major steps of cell respiration and the outcome of each step:

STAGES OF AEROBIC RESPIRATION				
Process	Location	Main Input	Main Output	Energy Output (per glucose)
Glycolysis	Cytoplasm	1 Glucose	2 pyruvates (2 net)	2 ATP 2 NADH
Formation of Acetyl-CoA	As pyruvate is transported into the mitochondria	2 pyruvates 2 Coenzyme A	2 Acetyl-CoA	2 NADH
Krebs or Citric Acid Cycle	Matrix of mitochondria	2 Acetyl-CoA oxaloacetate	oxaloacetate	6 NADH 2 $FADH_2$ 2 ATP
Oxidative Phosphorylation	Inner mitochondrial membrane	10 NADH 2 $FADH_2$ 2 O_2	Water 2 NAD^+ ($\times$ 1.5) 8 NAD^+ ($\times$ 2.5) 2 FAD^+ ($\times$ 1.5)	3 ATP $\rightarrow$ 20 ATP 3 ATP 4 ATP
				Overall Net: 30 ATP

Photosynthesis vs. Cell Respiration

The photosynthesis reactions share some similarities with cell respiration. In both cases, ATP production is driven by a proton gradient, and the proton gradient is created by an electron transport chain. In respiration, protons are pumped from the mitochondrial matrix to the intermembrane space, and they return to the matrix through an ATP synthase down their concentration gradient. In photosynthesis, protons are pumped from the stroma into the thylakoids compartment, and they return to the stroma through an ATP synthase down their concentration gradient.

The Krebs cycle and the Calvin cycle are both series of reactions that ultimately regenerate their starting product. Both cycles have an indirect need for a particular substance; although they do not use it directly, without it they cease to run. For the Krebs cycle, that substance is oxygen; for the Calvin cycle, that "substance" is light.

Furthermore, the goals of the two cycles are opposite; the Krebs cycle seeks to oxidize carbohydrates to CO_2, while the Calvin cycle seeks to reduce CO_2 to carbohydrates.

Anaerobic Respiration

When oxygen is not available, the anaerobic version of respiration occurs. The electron transport chain stops working, and electron carriers have nowhere to drop their electrons. The mitochondrial production of acetyl-CoA and the Krebs cycle cease too.

Glycolysis, however, can continue to run. This means that glucose can be broken down to give net two ATP. Only two instead of 30!

Glycolysis also gives two pyruvates and two NADH. The pyruvate and NADH make a deal with each other, and pyruvate helps NADH get recycled back into NAD^+ and takes its electrons. The pyruvate turns into either **lactic acid** (in muscles) or **ethanol** (in yeast). Since these two things are toxic at high concentrations, this process, called **fermentation**, is done only in emergencies. Aerobic respiration is a better option.

What types of organisms undergo fermentation?

- Yeast cells and some bacteria make ethanol and carbon dioxide.
- Other bacteria produce lactic acid.

Your Muscle Cells Can Ferment

Although human beings are aerobic organisms, they can actually carry out fermentation in their muscle cells. Have you ever had a leg cramp? If so, that cramp was possibly the consequence of anaerobic respiration.

When you exercise, your muscles require a lot of energy. To get this energy, they convert enormous amounts of glucose to ATP. But as you continue to exercise, your body doesn't get enough oxygen to keep up with the demand in your muscles. This creates an oxygen debt. What do your muscle cells do? They switch over to anaerobic respiration. Pyruvic acid produced from glycolysis is converted to lactic acid.

AP Biology Premium Prep

Scan the QR code below and register your book to access the digital flashcards of these formula terms:

CHAPTER 6 KEY TERMS

bioenergetics
First Law of Thermodynamics
Second Law of Thermodynamics
entropy
exergonic reaction
energy diagram
endergonic reaction
transition state
activation energy
enzymes
enzyme specificity
substrates
active site
enzyme-substrate complex
induced-fit
cofactors
coenzymes
denatured
Q_{10}
allosteric sites
competitive inhibition
allosteric inhibitor
noncompetitive inhibition
cellular respiration
photosynthesis
light reactions
dark reactions
photons
stroma
grana
thylakoids
chlorophyll *a* and *b*
carotenoids
reaction center
antenna pigments

photosystem I (PS I)
photosystem II (PS II)
P680
P700
photophosphorylation
absorption spectrum
emission spectrum
photolysis
NADPH
carbon fixation
Calvin-Benson Cycle
photorespiration
CAM plants
C_4 plants
aerobic respiration
anaerobic respiration
NADH
$FADH_2$
glycolysis
pyruvic acid
acetyl coenzyme A (acetyl CoA)
pyruvate dehydrogenase complex (PDC)
Krebs cycle (citric acid cycle)
matrix
oxaloacetate
citric acid
electron transport chain
cytochrome C
pH gradient (proton gradient)
chemiosmosis
ATP synthase
oxidative phosphorylation
lactic acid
ethanol (ethyl alcohol)
fermentation

AP Biology Premium Prep

Chapter 6 Drill

Answers and explanations can be found in Chapter 15.

1 ☐ Mark for Review

A solution of bromothymol blue indicator is exposed to carbon dioxide until the solution turns yellow. An aquatic plant, elodea, is placed into the yellow solution and exposed to the sun. After a few hours, the solution is now blue. Which of the following statements best describes the purpose of this procedure?

(A) The procedure proves that carbon dioxide is used up during photosynthesis.

(B) The procedure proves that oxygen is produced during photosynthesis.

(C) The procedure proves that carbon dioxide is produced during photosynthesis.

(D) The procedure proves that the bromothymol blue indicator becomes bromothymol when exposed to carbon dioxide.

2 ☐ Mark for Review

Carbon exists in multiple isotopes with similar chemical properties. Carbon labeling is used in chemical and biological experiments, which involves replacing regular carbon atoms with radioactive carbon-14 atoms that can be traced in chemical reactions. A group of plant samples is grown for several days in an environment containing CO_2 containing radioactive carbon-14 instead of nonradioactive CO_2. Which of the following statements describes the most likely result of this experiment?

(A) RuBP carboxylase will display increased activity in the plants' cells while they undergo photosynthesis.

(B) Radioactive water will be released from the plants' stomata.

(C) Radioactive carbon-14 will be found in the starch stored in the roots of the plants.

(D) Radioactive carbon-14 will be found in the leaves of the plants.

3 ☐ Mark for Review

A researcher adds carbonated water to a potted plant every day and notices that the plant is bigger and greener than other plants in the same location. Which of the following best explains how the carbonated water affected these results?

(A) The carbonated water is alkaline, which causes rapid growth in plants.

(B) The roots of the plant absorbed the O_2 from the carbonated water.

(C) The roots of the plant absorbed the CO_2 from the carbonated water.

(D) The leaves of the plant were exposed to more CO_2 and water from the carbonated water, both of which are required for photosynthesis.

4 ☐ Mark for Review

A botanist performs an experiment to explore where trees gain the raw materials needed to increase in mass. She plants a seedling weighing 3 pounds in a garden box containing 150 pounds of soil and regularly waters the seedling. After three years, the botanist weighs the tree and soil. The tree weighs 132 pounds, 5 ounces. The soil weighs 147 pounds, 10 ounces. Which of the following statements best explains these differences based on her data?

(A) The seedling's mass increased by incorporating soil from the garden box.

(B) The seedling's mass increased by incorporating carbon from CO_2 in the atmosphere.

(C) The seedling's mass increased by incorporating water from the botanist's regular watering.

(D) The seedling's mass increased by incorporating water from the botanist's regular watering and soil from the garden box.

6. Cellular Energetics | **161**

AP Biology Premium Prep

5 ⬚ Mark for Review

A cactus living in an arid desert has been performing photosynthesis throughout the day. Which of the following statements describes what will occur once the sun sets?

(A) Photosynthesis will no longer occur until the next morning.

(B) Oxygen will still be produced, but no carbon compounds will be produced until the next morning.

(C) Oxygen will be taken in through the stomata and stored to be used the next morning.

(D) Nothing will change since the Calvin cycle doesn't require sunlight and will continue to produce G3P.

6 ⬚ Mark for Review

Aerobic respiration, which converts energy from glucose to ATP, operates at approximately 40% efficiency. Which of the following statements best explains what happens to the remaining energy from glucose that is not converted?

(A) The remaining 60% of glucose's energy is stored in the mitochondrial membranes.

(B) The remaining 60% of glucose's energy is captured and utilized for anaerobic metabolic pathways.

(C) The remaining 60% of glucose's energy is expelled as heat energy and used to maintain homeostasis by keeping body temperatures consistent in warm-blooded organisms.

(D) The remaining 60% of glucose's energy is destroyed by enzymes catalyzing glycolysis reactions.

7 ⬚ Mark for Review

Glycolysis is a process involving ten specific reactions with ten intermediate compounds. The intermediates provide entry points to the overall process. Which of the following best describes the location where the process of glycolysis occurs?

(A) In the cytoplasm of both eukaryotic and prokaryotic cells

(B) In the folds of the membranes of prokaryotic cells

(C) In the membranes of both eukaryotic and prokaryotic cells

(D) In the mitochondrial matrix of eukaryotic cells

8 ⬚ Mark for Review

Some mammals hibernate to reduce their rate of cellular respiration during colder months. The rate of cellular respiration during hibernation is less than 50% of the normal cellular respiration rate. Which of the following characteristics likely occurs during cellular respiration?

(A) Increased heart rate and lower body temperature

(B) Increased heart rate and higher body temperature

(C) Decreased heart rate and lower body temperature

(D) Decreased heart rate and higher body temperature

162 | For more free content, visit PrincetonReview.com

AP Biology Premium Prep

9 | Mark for Review

Four cricket species are placed in a respirometer after being weighed. The rate of oxygen each species consumes is measured for 3 hours. The results of this research are shown below.

Species	Average Mass (mg)	Oxygen Consumed (cm^3)
A	150	0.03
B	75	0.015
C	50	0.01
D	25	0.005

According to the data, which of the following statements is the best conclusion the scientist can make from the data?

(A) The smallest cricket species consumed more than twice the oxygen per milligram of body weight than the largest cricket species consumed.

(B) All four species of crickets consumed the same amount of oxygen per milligram of body weight.

(C) The largest cricket species consumed more than twice the oxygen per milligram of body weight than the smallest cricket species consumed.

(D) Crickets placed in a colder respirometer will respire more rapidly than crickets placed in a warmer respirometer.

10 | Mark for Review

The rates of chemical reactions can be affected by adding a catalyst. Which of the following measurements can be used to mathematically determine the rate of chemical reactions with catalysts?

(A) The rate the substrate is used up

(B) The amount the activation energy increases

(C) The rate the product is used up

(D) The rate the enzyme is used up

6. Cellular Energetics | **163**

Chapter 6 Summary

- Energy cannot be created or destroyed. There are endergonic reactions (require energy) and exergonic reactions (release energy).
- The study of how cells accomplish biological processes is called bioenergetics.
  - Chemical reactions are catalyzed by enzymes.
  - Enzymes lower the activation energy of chemical reactions by stabilizing the transition state.
  - Enzymes do not change the energy difference between reactants and products, but by lowering the activation energy, they facilitate chemical reactions to occur.
- Enzymes are proteins that are highly specific for their substrate (or reactant, which binds at the active site).
- Enzymes have an optimal, narrow range of temperature and pH in which they have the highest rate of reaction. Outside this range, they may undergo denaturation and, therefore, no longer be active.
- Enzyme activity may also be regulated or altered by allosteric/noncompetitive inhibitors, competitive inhibitors, and activators.
- Enzymes may require coenzymes or cofactors to help catalyze reactions.
- ATP is the universal energy molecule in cells. It is created via cellular respiration.
- Photosynthesis is the process by which plants use energy from sunlight to make sugar.
  - In the light-dependent reactions, sunlight is absorbed by chlorophyll, electrons are passed down a chain, and the energy molecules ATP and NADPH are produced. Water is also split into hydrogen and oxygen.
  - In the light-independent reactions, sugar is made in the Calvin-Benson Cycle (Calvin cycle) by using energy from ATP and NADPH, and the carbon from CO_2.
- Cellular respiration consumes oxygen and produces a lot of ATP energy for the cell, using NADH and $FADH_2$ as electron carriers. There are four stages to cellular respiration:
  - glycolysis
  - formation of acetyl-CoA via the PDC
  - Krebs cycle
  - oxidative phosphorylation
- Anaerobic respiration occurs when cells lack oxygen to act as a final electron acceptor. In this process, the pyruvates generated by glycolysis are broken down by fermentation to produce lactic acid or ethanol and NAD^+, which permits glycolysis to continue and provide the 2 ATP it makes.

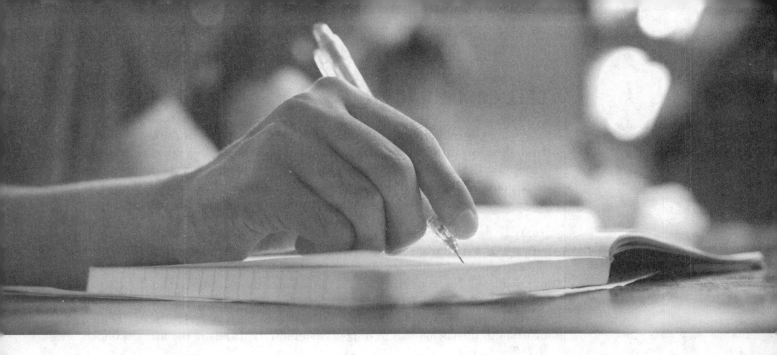

Chapter 7
Cell Communication and Cell Cycle

CELL COMMUNICATION

Communication is important for organisms and involves transduction of stimulatory or inhibitory signals from other cells, organisms, or the environment. Unicellular organisms detect and respond to environmental signals. For example, they can use cell communication to locate a suitable mate, adapt metabolism rates in response to changes in nutrient availability, or make their numbers known to other members of their species (a phenomenon known as **quorum sensing**).

Taxis is the movement of an organism in response to a stimulus and can be positive (toward the stimulus) or negative (away from the stimulus). Taxes are innate behavioral responses, or instincts. **Chemotaxis** is movement in response to chemicals. For example, bacteria can control flagella rotation to direct their motion, thus avoiding repellents (such as poisons) or helping them find favorable locations with high concentrations of attractants (such as food). In our bodies, neutrophils use chemotaxis to respond to an infection and are the first responders to inflammation.

Signal Transduction & Changes In Pathways

The cells of multicelled organisms must communicate with one another to coordinate the activities of the organism as a whole. Cells communicate through cell-to-cell contact or through cell signaling. Signaling can be short-range (affecting only nearby cells) or long-range (affecting cells throughout the organism). It can be done by cell junctions or signaling molecules called **ligands** that bind to **receptors** and trigger a response by changing the shape of the receptor protein. Hormones and neurotransmitters are common signaling molecules.

Signal transduction is the process by which an external signal is transmitted to the inside of a cell. It usually involves the following three steps:

1. a signaling molecule binding to a specific receptor
2. activation of a signal transduction pathway
3. production of a cellular response

Hydrophobic signaling molecules, like cholesterol-based steroid hormones, can diffuse across the plasma membrane. These molecules can bind receptors inside the cell and often travel to the nucleus where they regulate gene expression.

For signaling molecules that cannot enter the cell, a plasma membrane receptor is required. Plasma membrane receptors form an important class of integral membrane proteins that transmit signals from the extracellular space into the cytoplasm. Each receptor binds a particular molecule in a highly specific way.

There are three classes of membrane receptors.

1. **Ligand-gated ion channels** in the plasma membrane open or close an ion channel upon binding a particular ligand. An example is the ligand-gated sodium channel on the surface of a skeletal muscle cell at the neuromuscular junction. This channel opens in response to acetylcholine, and a massive influx of sodium depolarizes the muscle cell and causes it to contract.

2. **Catalytic (enzyme-linked) receptors** have an enzymatic active site on the cytoplasmic side of the membrane. Enzyme activity is initiated by ligand binding at the extracellular surface. The insulin receptor is an example of an enzyme-linked receptor. After binding insulin, enzymatic activity initiates a complex signaling pathway that allows the cell to grow, synthesize lipids, and import glucose.

3. A **G-protein-linked receptor** does not act as an enzyme, but instead will bind a different version of a G-protein (often GTP or GDP) on the intracellular side when a ligand is bound extracellularly. This causes activation of **secondary messengers** within the cell. One important second messenger is cyclic AMP (cAMP). It is known as a "universal hunger signal" because it is the second messenger of the hormones epinephrine and glucagon, which cause energy mobilization. Secondary messengers are usually small molecules that can move quickly through the cell. They can be made and destroyed quickly and help the signal amplify throughout the cell.

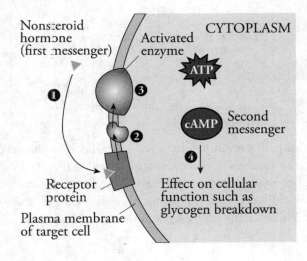

Signal transduction in eukaryotic cells usually involves many steps and complex regulation. Many signal transduction pathways include protein modifications and phosphorylation cascades, in which a series of protein kinases add a phosphate group to the next protein in the cascade sequence. In contrast, bacterial cells usually use a simpler two-component regulatory system in transduction pathways.

Signal transduction cascades are helpful to amplify a signal. They are also helpful in quickly turning a response on or off and can lead to important changes in gene expression and cell function.

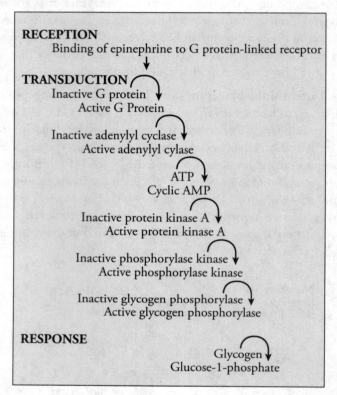

Signal transduction pathways influence how cells respond to their environments. Pathways may cause changes to gene expression and cell function, which may change a cell's phenotype or cause apoptosis. Mutations in the receptor or the ligand can alter signaling. Drugs and chemicals that inhibit or excite part of the pathway can also influence the gene expression within the cell.

FEEDBACK

All living things have a "steady state" that they try to maintain. This is helpful because the structures and processes within living things are sensitive to things like temperature, pH, pressure, salinity, osmotic pressure, and many other things. The set of conditions under which living things can successfully survive is called **homeostasis**. The body is constantly working to maintain this state by taking measurements and then responding appropriately. That might mean shivering if it is cold or releasing the hormone insulin if blood glucose is high. Depending on the condition, the body will respond accordingly.

Many of these responses are controlled by negative or positive feedback pathways. A **negative feedback pathway** (also called **feedback inhibition**) works by turning itself off using the end product of the pathway. The end product inhibits the process from beginning, thus shutting down the pathway. This is a common strategy to conserve energy. If plenty of X is made, then the pathway to make X can be turned off. Sometimes, a very distant end product will turn off a process occurring several steps back. For example, ATP turns off the beginning stages of glycolysis, although ATP is not made until much later in cellular respiration. There are several other types of second messengers, such as cyclic GMP, calcium ions, and inositol triphosphate.

A **positive feedback pathway** also involves an end product playing a role, but instead of inhibiting the pathway, it further stimulates it. For example, once X is made, it tells the pathway to make even more X. This is less common, but it occurs during fruit ripening and labor and delivery, which ramps up and up as it proceeds.

Your blood glucose levels are regulated by insulin and glucagon, two hormones released from your pancreas. For example, when you've just eaten, blood sugar levels are high, and this stimulates insulin release. The cells of your body take up glucose and hepatocytes in your liver and store extra glucose as glycogen. Blood sugar levels go down, your body reestablishes homeostasis, and insulin release is shut off via negative feedback.

Negative

Positive

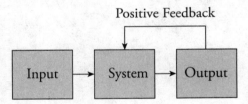

You won't need to remember this specific example, but if it helps to better visualize the positive feedback pathway, consider this. When a baby pushes their head against the mother's womb, the womb's stretching causes a hormone called oxytocin to be released. In turn, oxytocin triggers the mother's womb muscles to contract, pushing the baby harder against the womb. This causes a cycle: further stretching of the womb, further oxytocin release, more womb contractions, and so on, until the baby is delivered.

Cell Communication in Plants

Plants do not have a nervous system but can produce several proteins found in animal nervous systems, such as certain neurotransmitter receptors. Plants can also generate electrical signals in response to environmental stimuli, and this can affect flowering, respiration, photosynthesis, and wound healing. Light receptors are common in plants and help link environmental cues to biological processes such as seed germination, the timing of flowering, and chlorophyll production. Some plants can also use chemicals to communicate with nearby plants. For example, wounded tomatoes produce a volatile chemical as an alarm signal. This warns nearby plants and allows them to prepare a defense or immune response.

THE CELL CYCLE

Every second, thousands of cells are dying throughout our bodies. Fortunately, the body replaces them at an amazing rate. In fact, epidermal cells, or skin cells, die off and are replaced so quickly that the average 18-year-old grows an entirely new skin every few weeks. The body keeps up this unbelievable rate thanks to the mechanisms of **cell division**.

The remainder of this chapter takes a closer look at how cells divide. But remember, cell division is only a small part of the life cycle of a cell. Most of the time, cells are busy carrying out their regular activities. There are also some types of cells that are nondividing. Often, these are highly specialized cells that are created from a population of less specialized cells. The body continually makes them as needed, but they do not directly replicate themselves. Red blood cells are an example of nondividing cells. Some other cells are just temporarily nondividing. They enter a phase called G_0, where they hang out until they get a signal to reenter the normal cell cycle.

Every cell has a life cycle—the period from the beginning of one division to the beginning of the next. The cell's life cycle is known as the **cell cycle**. The cell cycle is divided into two periods: **interphase** and **mitosis**. Take a look at the cell cycle of a typical cell.

Special gametes split their time between interphase and meiosis.

Notice that most of the life of a cell is spent in interphase.

INTERPHASE: THE GROWING PHASE

Interphase is the time span from one cell division to another. We call this stage interphase (*inter-* means between) because the cell has not yet started to divide. Although biologists sometimes refer to interphase as the "resting stage," the cell is definitely not inactive. During this phase, the cell carries out its regular activities. All the proteins and enzymes it needs to grow are produced during interphase.

The Three Stages of Interphase
Interphase can be divided into three stages: **G₁, S, G₂**.

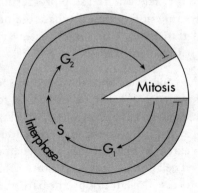

The most important phase is the S phase. That's when the cell replicates its genetic material. The first thing a cell has to do before undergoing mitosis is to duplicate all of its chromosomes, which contain the organism's DNA "blueprint." During interphase, every single chromosome in the nucleus is duplicated.

You'll notice that the original chromosome and its duplicate are still linked, like conjoined twins. These identical strands of DNA are now called **sister chromatids**. The chromatids are held together by a structure called the **centromere**. You can think of each chromatid as a chromosome, but because they remain attached, they are called chromatids instead. To be called a chromosome, each needs to have its own centromere. Once the chromatids separate, they will be full-fledged chromosomes.

Cell Cycle Regulation
We've already said that replication occurs during the S phase of interphase, so what happens during G₁ and G₂? During these stages, the cell performs metabolic reactions and produces organelles, proteins, and enzymes. For example, during G₁, the cell produces all of the enzymes required for DNA replication (DNA helicase, DNA polymerase, and DNA ligase, which are explained later in Chapter 9). By the way, *G* stands for "gap," but we can also associate it with "growth." These three phases are highly regulated by checkpoints and special proteins called **cyclins** and **cyclin-dependent kinases (CDKs)**.

Cell cycle checkpoints are control mechanisms that make sure cell division is happening properly in eukaryotic cells. They monitor the cell to make sure it is ready to progress through the cell cycle and stop progression if it is not. In eukaryotes, checkpoint pathways function mainly at phase boundaries (such as the G_1/S transition and the G_2/M transition). Checkpoint pathways are an example of an important cell signaling pathway inside cells.

For example, if the DNA genome has been damaged in some way, the cell should not divide. If it does, damaged DNA will be passed on to daughter cells, which can be disastrous. When damaged DNA is found, checkpoints are activated and cell cycle progression stops. The cell uses the extra time to repair damage in DNA. If the DNA damage is so extensive that it cannot be repaired, the cell can undergo **apoptosis**, or programmed cell death. Apoptosis cannot stop once it has begun, so it is a highly regulated process. It is an important part of normal cell turnover in multicellular organisms.

Cell cycle checkpoints control cell cycle progression by regulating two families of proteins: cyclin-dependent kinases (CDKs) and cyclins. To induce cell cycle progression, an inactive CDK binds a regulatory cyclin. Once together, the complex is activated, can affect many proteins in the cell, and causes the cell cycle to continue. To inhibit cell cycle progression, CDKs and cyclins are kept separate. CDKs and cyclins were first studied in yeast, unicellular eukaryotic fungi. Budding yeast and fission yeast were good model organisms because they have only one CDK protein each and a few different cyclins. This relatively simple system allowed biologists to figure out how cell cycle progression is controlled. Their findings were then used to figure out how more complex eukaryotes (such as mammals) control the cell cycle.

Cancer

A cell losing control of the cell cycle can have disastrous consequences. For example, a mutation in a protein that normally controls progression through the cell cycle can result in unregulated cell division and cancer. **Cancer** means "crab," as in the zodiac sign. The name comes from the observation that malignant tumors grow into surrounding tissue, embedding themselves like clawed crabs. Cancer occurs when normal cells start behaving and growing very abnormally and spread to other parts of the body.

Mutated genes that induce cancer are called **oncogenes** ("onco-" is a prefix denoting cancer). Normally, these genes are required for proper growth of the cell and regulation of the cell cycle. Oncogenes, then, are genes that can convert normal cells into cancerous cells. Often these are abnormal or mutated versions of normal genes. The normal healthy version is called a proto-oncogene.

Tumor suppressor genes produce proteins that prevent the conversion of normal cells into cancer cells. They can detect damage to the cell and work with CDK/cyclin complexes to stop cell growth until the damage can be repaired. They can also trigger apoptosis if the damage is too severe to be repaired.

Cell cycle checkpoints and apoptosis have been intimately linked with cancer. In order for a normal cell to become a cancer cell, it must override cell cycle

Both oncogenes and tumor suppressors can cause cancer, but oncogenes do it by stepping on the cell cycle gas pedal, while tumor suppressors do it by removing the brake pedal.

checkpoints and grow in an unregulated way. It must also avoid cell death. Tumors often accumulate DNA damage. Cancer treatments target these changes, because they are what make a cancer cell different from a normal cell.

Let's recap:

- The cell cycle consists of two things: interphase and mitosis.
- During the S phase of interphase, the chromosomes replicate via DNA replication.
- Growth and preparation for mitosis occur during the G_1 and G_2 stages of interphase.
- Cell cycle checkpoints make sure the cell is ready to continue through the cell cycle.
- CDK and cyclin proteins work together to promote cell cycle progression.
- Oncogenes promote cell growth and tumor suppressor genes inhibit cell growth.

MITOSIS: THE DANCE OF THE CHROMOSOMES

Once the chromosomes have replicated, the cell is ready to begin mitosis—the period during which the cell divides into two genetically identical daughter cells. Mitosis is important for growth, tissue repair, and asexual reproduction. Mitosis consists of a sequence of four stages: **prophase**, **metaphase**, **anaphase**, and **telophase**. It is not important to memorize the name of each phase, but it is important that you know the basic order of operations for mitosis.

Remember, the genome is all the DNA, or all the chromosomes, in a cell.

Stage 1: Prophase: Prep (the cell prepares to divide)

One of the first signs of prophase is the disappearance of the nucleolus (ribosome-making area) and the nuclear envelope. In addition, the genome becomes visible as individual chromosomes; diffuse **chromatin** (which looks like a pile of scrunched up dental floss in the nucleus) condenses into densely packed chromosomes. To accomplish this, chromatin forms coils upon coils and thickens. If you look at a human cell under the light microscope at the beginning of prophase, you would see 46 differently sized chromosomes, and each one would be the shape of an X. During interphase, individual chromosomes are not visible.

Condensed chromosomes are easier to separate and move around the cell compared to diffuse chromatin.

Reminder: Plants usually don't have centrioles, but microtubules are still used to pull the chromatids apart.

Another name for this MTOC in animal cells is the centrosome.

Now the cell has plenty of room to "sort out" the chromosomes. Remember centrioles? During prophase, these cylindrical bodies found within microtubule organizing centers (MTOCs) start to move away from each other, toward opposite ends of the cell. The centrioles will spin out a system of microtubules known as the **spindle fibers**. These spindle fibers will attach to a **kinetochore**, a structure on each chromatid. The kinetochores are part of the centromere.

Stage 2: Metaphase: Middle (chromosomes align in the middle)

The next stage is called metaphase. The chromosomes now begin to line up along the equatorial plane, or the **metaphase plate**, of the cell. This happens because spindle fibers are attached to the kinetochore of each chromatid.

Metaphase plate

Stage 3: Anaphase: Apart (chromatids are pulled apart)

During anaphase, sister chromatids of each chromosome separate at the centromere and migrate to opposite poles. Chromatids are pulled apart by the microtubules, which begin to shorten. Each half of a pair of sister chromatids now moves to opposite poles of the cell. Non-kinetochore microtubules elongate the cell.

Anaphase

Stage 4: Telophase: Two (the cell completes splitting in two)

The final phase of mitosis is telophase. A nuclear membrane forms around each set of chromosomes and the nucleoli reappear.

The nuclear membrane is ready to divide. Now it's time to split the cytoplasm in a process known as **cytokinesis**. Look at the figure below and you'll notice that the cell has begun to split along a **cleavage furrow** (which is produced by actin microfilaments).

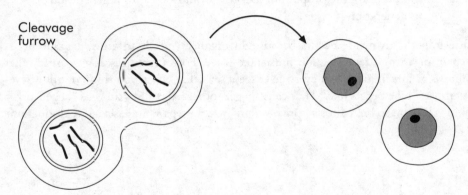

A cell membrane forms around each cell, and the cell splits into two distinct daughter cells. The division of the cytoplasm yields two daughter cells.

Here's one thing to remember: cytokinesis occurs differently in plant cells. The cell doesn't form a cleavage furrow. Instead, a partition called a **cell plate** forms down the middle region.

Stage 5: Interphase

Once daughter cells are produced, they reenter the initial phase—interphase—and the whole process starts over. The cell goes back to its original state. Once again, the chromosomes decondense and become invisible, and the genetic material is called chromatin again.

But How Will I Remember All That?

For mitosis, you may already have your own mnemonic. If not, here's a table with a mnemonic we created for you.

IPMAT	
Interphase	**I** is for **Interlude**
Prophase	**P** is for **Prepare**
Metaphase	**M** is for **Middle**
Anaphase	**A** is for **Apart**
Telophase	**T** is for **Two**

Purpose of Mitosis

Mitosis achieves two things:

- the production of daughter cells that are identical copies of the parent cell
- maintaining the proper number of chromosomes from generation to generation

The impetus to divide occurs because an organism needs to grow, a tissue needs repair, or asexual reproduction must take place. For our purposes, we can say that mitosis occurs in just about every cell except sex cells. When you think of mitosis, remember: "Like begets like." Hair cells "beget" other hair cells, skin cells "beget" other skin cells, and so on. Mitosis is involved in growth, repair, and asexual reproduction.

CHAPTER 7 KEY TERMS

quorum sensing
taxis
chemotaxis
ligands
receptors
signal transduction
ligand-gated ion channel
catalytic (or enzyme-linked) receptor
G-protein-linked
  receptor
secondary messengers
homeostasis
negative feedback
  pathway (feedback inhibition)
positive feedback
  pathway
cell division
G_0 phase
cell cycle
interphase
mitosis
G_1 phase

S phase
G_2 phase
sister chromatids
centromere
cyclins
cyclin-dependent kinases (CDKs)
cell cycle checkpoints
apoptosis
cancer
oncogene
tumor suppressor gene
prophase
metaphase
anaphase
telophase
chromatin
spindle fibers
kinetochores
metaphase plate
cytokinesis
cleavage furrow
cell plate

Chapter 7 Drill

Answers and explanations can be found in Chapter 15.

1 ▢ Mark for Review

Signal transduction pathways sometimes require multiple steps to achieve a certain response. What is the advantage of having many steps in one pathway?

(A) The enzymes for each step are often located relatively far apart in the cell, so having multiple steps provides time for them to reach each other.

(B) Having multiple steps allows the response to be amplified, triggering a large response from one signal molecule.

(C) By breaking the task into multiple steps, the cell can release energy in small increments instead of overheating the cell with one large release of heat.

(D) The pathway can produce more products by repeating the same steps multiple times

2 ▢ Mark for Review

Adrenaline is a neurotransmitter and a hormone that plays a significant role in the human body's flight-or-fight response. When adrenaline binds to a receptor on smooth muscle, the smooth muscle relaxes. When adrenaline binds to a receptor on skeletal muscle, the skeletal muscle contracts. What causes the different reactions in the two types of muscle when exposed to adrenaline?

(A) A different signal transduction pathway is triggered by the binding of adrenaline to receptors on smooth muscle when compared to skeletal muscle, which results in the two cell types experiencing different responses.

(B) The smooth muscle cells are permeable to adrenaline, which activates different genes through transcription factors.

(C) The smooth muscle cells do not contain any secondary messengers, such as cGMP or IP_3.

(D) Adrenaline binds to a tyrosine kinase receptor on one cell type and a G protein-coupled receptor on another cell type.

3 ▢ Mark for Review

Cancer is an extensive category of diseases caused by unregulated cell growth, where cells grow uncontrollably and form malignant tumors that invade nearby and distant parts of the body. Some tumors are benign, which do not grow uncontrollably or invade other areas of the body. Which of the following identifies a characteristic of cancer cells when compared to healthy cells?

(A) Cell division is more carefully managed.

(B) The rate of mitosis is lower.

(C) Protein translation occurs at a higher rate.

(D) The cell cycle is shorter.

AP Biology Premium Prep

4 ☐ Mark for Review

Which of the following processes occurs during traditional mitosis?

(A) Chromosomes pair up in homologous pairs.

(B) Chromosomes from paternal gametes line up with maternal gametes.

(C) Chromosomes line up in the middle of the cell before they're replicated.

(D) Chromosomes line up in the middle of the cell once they're replicated.

5 ☐ Mark for Review

Which of the following correctly pairs the structure and function of part of a chromosome in a eukaryotic cell?

(A) Centromeres help attach the mitotic spindle.

(B) Nucleosomes hold sister chromatids together.

(C) Chromatids distribute one copy of each chromosome to daughter cells.

(D) Kinetochores help package DNA into a compact structure.

6 ☐ Mark for Review

Which of the following mutations would be most likely to have a discernible change in the phenotype of the individual?

(A) Deletion of 400 base pairs resulting in the loss of an enhancer

(B) Translocation of the last 1,000 base pairs of chromosome 1 onto chromosome 2

(C) Translocation of the first 1,000 base pairs of chromosome 2 onto chromosome 3

(D) Substitution of a single base that results in the same amino acid as the original codon

7. Cell Communication and Cell Cycle | **179**

Chapter 7 Summary

- Sometimes cells do not need to pass any molecules through the membrane, but instead they just pass a message. This is signal transduction. A ligand binds to a receptor on the outside of the cell and causes changes to the inside of the cell. Ligand-gated ion channels, catalytic receptors, and G-protein-linked receptors are common examples.

- The cell cycle is divided into interphase and mitosis, or cellular division.

- The three stages of interphase are G_1, G_2, and S phase.
  - S phase is the "synthesis" phase, when chromosomes replicate.
  - Growth and preparation for mitosis occur in G_1 and G_2.

- Cell cycle progression is controlled by checkpoint pathways and CDK/cyclin complexes.

- Cancer occurs when cells grow abnormally and spread to other parts of the body. Tumor-suppressor genes are genes that prevent the cell from dividing when it shouldn't. Proto-oncogenes are genes that help the cell divide. If either type is mutated (a mutated proto-oncogene is called an oncogene), it can lead to cell growth that is out of control.

- Mitosis, or cellular division, occurs in four stages: prophase, metaphase, anaphase, and telophase.
  - During prophase, the nuclear envelope disappears and chromosomes condense.
  - Next is metaphase, when chromosomes align at the metaphase plate and mitotic spindles attach to kinetochores.
  - In anaphase, chromosomes are pulled away from the center.
  - Telophase terminates mitosis, and the two new nuclei form.
  - The process of cytokinesis, which occurs during telophase, ends mitosis, as the cytoplasm and plasma membranes pinch to form two distinct, identical daughter cells.

Chapter 8
Heredity

HAPLOIDS VERSUS DIPLOIDS

Every organism has a certain number of unique chromosomes. For example, fruit flies have 4 chromosomes, humans have 23 chromosomes, and dogs have 39 chromosomes. It turns out that most eukaryotic cells in fact have two full sets of chromosomes—one set from each parent. Humans, for example, have two sets of 23 chromosomes, giving us our grand total of 46.

A cell that has two sets of chromosomes is a **diploid cell**, and the zygotic chromosome number is given as "$2n$." That means we have two copies of each chromosome.

If a cell has only one set of chromosomes, we call it a **haploid cell**. This kind of cell is given the symbol n. For example, we would say that the haploid number of chromosomes for humans is 23.

Remember:

- *Diploid* refers to any cell that has two sets of chromosomes.
- *Haploid* refers to any cell that has one set of chromosomes.

Why do we need to know the terms *haploid* and *diploid*? Because they are extremely important when it comes to sexual reproduction. As we've seen, 46 is the normal diploid number for human beings, but there are only 23 different chromosomes. The duplicate versions of each chromosome are called **homologous chromosomes**. The homologous chromosomes that make up each pair are similar in size and shape and contain the same genes in the same locations (although they might have different versions of those genes). This is the case in all sexually reproducing organisms. In fact, this is the essence of sexual reproduction: each parent donates half its chromosomes to its offspring.

> Remember that the *n* stands for the number of unique chromosomes.

Gametes

Although most cells in the human body are diploid, there are special cells that are haploid. These haploid cells are called **sex cells**, or **gametes**. Why do we have haploid cells?

As we've said, an offspring has one set of chromosomes from each of its parents. A parent, therefore, contributes a gamete with one set that will be paired with the set from the other parent to produce a new diploid cell, or zygote. This produces offspring that are a combination of their parents.

GREGOR MENDEL: THE FATHER OF GENETICS

What is genetics? In its simplest form, genetics is the study of heredity. It explains how certain characteristics are passed on from parents to children. Much of what we know about genetics was discovered by the monk **Gregor Mendel** in the 19th century. Since then, the field of genetics has vastly expanded.

Let's begin then with some of the fundamental points of genetics:

- **Traits**—or expressed characteristics—are influenced by one or more of your genes. Remember, a **gene** is simply a chunk of DNA that codes for a particular recipe. Different recipes affect different traits. DNA is passed from generation to generation, and from this process the genes and the traits associated with those genes are inherited. Within a chromosome, there are many genes, and each can contribute to a different trait. For example, in pea plants, there's a gene on a chromosome that codes for seed coat characteristics. The position of a gene on a chromosome is called a **locus**.

- Diploid organisms (organisms that have two sets of chromosomes) usually have two copies of each gene, one on each homologous chromosome. Homologous chromosomes are two copies or versions of the same chromosome in a diploid cell or organism. Humans have 23 pairs of homologous chromosomes.

  The two gene copies may be different from one another—that is, they may be alternate versions, or **alleles**, of the same gene. For example, if we're talking about the height of a pea plant, there's an allele for tall and an allele for short. In other words, both alleles are alternate forms of the gene for height.

- When an organism has two identical alleles for a given trait, the organism is **homozygous**. If an organism has two *different* alleles for a given trait, the organism is **heterozygous**.

- When discussing the physical appearance of an organism, we refer to its **phenotype**. The phenotype tells us what the organism looks like. When talking about the genetic makeup of an organism, we refer to its **genotype**. The genotype tells us which alleles the organism possesses.

- An allele can be **dominant** or **recessive**. This is determined by which allele "wins out" over the other in a heterozygote. The convention is to assign one of two letters for the two different alleles. The dominant allele receives a capital letter, and the recessive allele receives a lowercase of the same letter. For instance, we might give the dominant allele for height in pea plants a *T* for tall. This means that the recessive allele would be *t* for short.

- Major components of the genetic code are shared by all modern living organisms, including ribosomes and nucleic acids (DNA/RNA). Many core metabolic pathways are conserved across all domains, which provides evidence for a common ancestry for all living organisms.

Overall then, homologous chromosomes are the same size and shape, and contain the same genes. However, they can contain different versions (or alleles) of those genes, and thus have different genetic sequences.

Name	Genotype	Phenotype
homozygous dominant	*TT*	Tall
homozygous recessive	*tt*	Short
heterozygous	*Tt*	Tall

One of the major ways the testing board likes to test genetics is by having you do crosses. Crosses involve mating hypothetical organisms with specific phenotypes and genotypes. We'll look at some examples in a moment, but for now, keep these test-taking tips in mind:

- Label each generation in the cross. The first generation in an experiment is always called the **parent**, or **P generation**. The offspring of the P generation are called the first **filial**, or F_1 **generation**. Members of the next generation, the grandchildren, are called the F_2 **generation**.
- Always write down the symbols you're using for alleles, along with a key to remind yourself what the symbols refer to. Use uppercase for dominant alleles and lowercase for recessive alleles.

Now let's look at some basic genetic principles.

One of Mendel's hobbies was to study the effects of cross-breeding on different strains of pea plants. Mendel worked exclusively with true-breeder pea plants. This means the plants he used were genetically pure and consistently produced the same traits. For example, tall plants always produced tall plants; short plants always produced short plants. Through his work, he came up with three principles of genetics: the **Law of Dominance**, the **Law of Segregation**, and the **Law of Independent Assortment**.

The Law of Dominance
Mendel crossed two true-breeding plants with contrasting traits: tall pea plants and short pea plants.

To his surprise, when Mendel mated these plants, the characteristics didn't blend to produce plants of average height. Instead, all the offspring were tall.

Mendel recognized that one trait must be masking the effect of the other trait. This is called the Law of Dominance. The dominant tall allele, T, somehow masked the presence of the recessive short allele, t. Consequently, all a plant needs is one tall allele to make it tall.

Monohybrid Cross

A **monohybrid cross** occurs when two individuals are crossed and one gene is being studied. A simple way to represent a monohybrid cross is to set up a **Punnett square**. Punnett squares are used to predict the results of a cross. Let's construct a Punnett square for the cross between Mendel's true-breeding tall and short pea plants. Let's first designate the alleles for each plant. As we saw earlier, we can use the letter T for the tall dominant allele and t for the recessive short allele.

Since one parent was a pure, tall pea plant, we'll give it two dominant alleles (TT homozygous dominant). The other parent was a pure, short pea plant, so we'll give it two recessive alleles (tt homozygous recessive). To look at the types of offspring

Know Your Way Around a Square

A Punnett square is just a visual way to see all the possible offspring combos that could result from two parents mating. The possible gametes for each parent go on the sides and then they get matched up with each other to show the possible offspring genotypes.

Remember, these squares are looking only at one (or sometimes two) genes. So, they show possible offspring genotypes for that one gene only.

that these parents could produce, we have to figure out the gametes that each parent can possibly make (and thus contribute to the offspring).

Each parent has two copies of the height gene and each gamete gets one copy. This means each parent can make two types of gametes. For the homozygous dominant parent, the two copies are *T* and *T*. In other words, all gametes from this parent will contain the tall, or *T*, allele. The homozygous recessive plant can make gametes with *t* and with *t*. Notice again that because the parent plant was homozygous, only one type of allele (*t*, or short) will be passed on to the next generation. We will put the possible gametes for each parent on the top and left side of the square as shown below.

Keep This In Mind
When a parent is homozygous, both of their gametes will be identical, but it is still helpful to list each gamete individually on the square.

Now we can fill in the four boxes by matching the letters. What are the results for the F_1 generation?

	T	T
t	Tt	Tt
t	Tt	Tt

Each offspring received one allele from each parent. They all received one *T* and one *t*. They're all *Tt*! So, even though the parents were homozygous, offspring are heterozygous: they possess one copy of each allele.

The Law of Segregation

Next, Mendel took the offspring and self-pollinated them. Let's use a Punnett square to spell out the results. This time we're mating the offspring of the first generation—F_1; we are crossing those heterozygotes. Take a look at the results:

	T	t
T	TT	Tt
t	Tt	tt

F_2 Generation

Check out the lower right box. One of the offspring could be a short pea plant!

Although all of the F_1 plants were tall, they were heterozygotes that could make gametes containing the short *t* allele. In other words, their *T* and *t* alleles were segregated (separated) during gamete formation. This is an example of the Law of Segregation. It just means that each gamete gets only one of two copies of a gene. Gametes are haploid.

What about the genotype and phenotype for this cross? Remember, genotype refers to the genetic makeup of an organism, whereas phenotype refers to the appearance of the organism. According to the results of our Punnett square, what is the ratio of phenotypes and genotypes in the offspring?

Let's sum up the results. We have four offspring with two different phenotypes: three of the offspring are tall, whereas one of them is short. On the other hand, we have three genotypes: 1 *TT*, 2 *Tt*, and 1 *tt*.

Here's a summary of the results:

- The ratio of phenotypes is 3:1 (three tall:one short).
- The ratio of genotypes is 1:2:1 (one *TT*:two *Tt*:one *tt*).

The Law of Independent Assortment

So far, we have looked at only one trait: tall versus short. What happens when we study two traits at the same time? Each allele of the two traits will get segregated into two gametes, but how one trait gets split up into gametes has no bearing on how the other trait gets split up. In other words, the two alleles for each trait are sorted independently of the two alleles for other traits. For example, let's look at two traits in pea plants: height and color. When it comes to height, a pea plant can be either tall (*T*) or short (*t*). As for color, the plant can be either green (*G*) or yellow (*g*), with green being dominant. Each gene comes in two alleles, so this gives us four alleles total. By the Law of Independent Assortment, these four alleles can combine to give us four different gametes.

$$TG \quad Tg \quad tG \quad tg$$

In other words, each allele of the height gene can be paired with either allele of the color gene. The reason alleles segregate independently is because chromosomes segregate independently. During meiosis I, each pair of homologous chromosomes is split, and which one aligns left or right during metaphase I is different for each pair.

Dihybrid Cross

Now that we know that different genes assort independently into gametes, let's look at a cross of two traits using those four gametes that we created above. A **dihybrid cross** is just like the monohybrid, but it studies how two genes are passed on to offspring.

Here is the Punnett Square for a cross between two double heterozygotes (*Tt Gg*).

	TG	Tg	tG	tg
TG	TTGG	TTGg	TtGG	TtGg
Tg	TTGg	TTgg	TtGg	Ttgg
tG	TtGG	TtGg	ttGG	ttGg
tg	TtGg	Ttgg	ttGg	ttgg

This is an example of the Law of Independent Assortment. Each of the traits segregated independently. Don't worry about the different combinations in the cross—you'll make yourself dizzy with all those letters. Simply memorize the phenotype ratio of the pea plants. For the 16 offspring there are:

- 9 tall and green
- 3 tall and yellow
- 3 short and green
- 1 short and yellow

That's 9:3:3:1, but don't forget the circumstances to which this applies. It is only when two heterozygotes for two genes are crossed. It is not just a magic ratio that always works.

Rules of Probability

The Punnett square method works well for monohybrid crosses and helps us visualize possible combinations. However, a better method for predicting the likelihood of certain results from a dihybrid cross is to apply the Rules of Probability. To determine the probability that two or more independent events will occur simultaneously, one can simply calculate the product of the probability that each will occur independently. This is called the **Product Rule**. To determine the likelihood that EITHER event occurs, but not both, use another rule called the **Sum Rule**. To illustrate the product rule, let's consider again the cross between two identical dihybrid tall, green plants with the genotype $TtGg$.

> Product Rule: If A and B are independent, then:
> P(A and B) = P(A) times P(B)
>
> Sum Rule: If A and B are mutually exclusive, then:
> P(A or B) = P(A) + P(B)

> Probability can be expressed as a fraction, percentage, or a decimal. Remember that this rule works only if the results of one cross are not affected by the results of another cross.

To find the probability of having a tall, yellow plant, simply multiply the probabilities of each event. If the probability of being tall is $\frac{3}{4}$ and the probability of being yellow is $\frac{1}{4}$, then the probability of being tall and yellow is $\frac{3}{4} \times \frac{1}{4} = \frac{3}{16}$.

Let's summarize Mendel's three laws.

SUMMARY OF MENDEL'S LAWS	
Laws	**Definition**
Law of Dominance	One trait masks the effects of another trait.
Law of Segregation	Each gamete gets only one of the copies of each gene.
Law of Independent Assortment	Each pair of homologous chromosomes splits independently, so the alleles of different genes can mix and match.

Test Cross

Suppose we want to know if a tall plant is homozygous (TT) or heterozygous (Tt). Its physical appearance doesn't necessarily tell us about its genetic makeup. The only way to determine its genotype is to cross the plant with a recessive, short plant, tt. This is known as a **test cross**. When the recessive plant is used, there will be only two possible outcomes in the daughter plants. Our mystery tall plant could be TT or Tt so when we mate it to a short plant, these are the possible situations: (1) $TT \times tt$ or (2) $Tt \times tt$. Let's take a look.

If none of the offspring is short, our original plant must have been homozygous, TT. If, however, even one short plant appears in the bunch, we know that our original pea plant was heterozygous, Tt. In other words, it wasn't a pure-breeding plant. Overall then, a test cross uses a recessive organism to determine the genotype of an organism of unknown genotype.

NON-MENDELIAN GENETICS

Some traits do not follow traditional patterns of inheritance or the ratios predicted by Mendel's laws; observed phenotypic ratios differ statistically from predicted rates.

Linked Genes

Sometimes genes on the *same* chromosome stay together during assortment and move as a group. The group of genes is considered linked and tends to be inherited together. For example, the genes for flower color and pollen shape are linked on the same chromosomes and show up together.

Since **linked genes** are found on the same chromosome, they cannot segregate independently. This violates the Law of Independent Assortment.

Let's pretend that the height and color genes we looked at in the dihybrid cross are linked. A heterozygote for both traits still has two alleles for height (*T* and *t*) and two alleles for color (*G* and *g*). However, because height and color are located on the same chromosome, one allele for height and one allele for color are physically linked. For example, maybe the heterozygote has one chromosome with *Tg* and one chromosome with *tG*. When gametes are formed, the *T* and *g* will travel together, and *t* and *G* will travel together and be packaged into a gamete together. So, in the unlinked dihybrid shown earlier, there were four possible gamete combinations (*TG*, *Tg*, *tG*, *tg*), but now there are only two (*Tg* and *tG*). The only way to physically separate linked alleles is by crossing-over. If a crossover event occurs between linked genes, then recombinant gametes can occur. However, because linked genes are physically close together, crossing-over is not likely to occur between them.

If the genes were unlinked, then four gametes (*TG*, *Tg*, *tG*, *tg*) would be equally likely (as we saw when we talked about independent assortment). However, if certain combinations of alleles are found more often in offspring than they should (based on Mendelian ratios), this is a red flag that the two genes are close together and linked.

Look at the example below from a cross of *TtGg* and *ttgg*. If you draw out the Punnett square, you should get equal numbers of each phenotype occurring if the genes are assumed to be unlinked. Yet, what if the numbers below were the actual numbers of offspring of each phenotype?

Tall and Green: 6 offspring

short and Green: 37 offspring

Tall and yellow: 45 offspring

short and yellow: 3 offspring

There are more Green/short and yellow/Tall offspring. This is a sign that these alleles are linked on a chromosome because they occur more often than they statistically should. This shows that most of the gametes of the *TtGg* parent had *tG* together and *Tg* together since those gamete combinations would result in Green/short and yellow/Tall offspring.

So, where did the Green/Tall and yellow/short plants come from? They are the result of gametes from the *TtGg* parent that are *TG* and *tg*, respectively. If the *Tg* and *tG* alleles were linked in that parent, how could they produce a few *TG* and *tg* gametes? The answer is through recombination! Even though those alleles were not found on the same chromosome, recombination must have occurred during meiosis to swap them. This allowed gametes to be formed with *TG* and *tg*, and offspring to occur that were Tall/Green and short/yellow. The offspring formed from recombination events are called **recombinants**. The **percentage of recombination (recombination frequency)** can be determined by adding up the recombinants and dividing by the total number of offspring.

$$9/91 = 9.9\%$$

This percentage can also be used as a measure of how far apart the genes are. The distance on a chromosome is measured in **map units**, or **centimorgans**.

The frequency of crossing-over between any two linked alleles is proportional to the distance between them. That is, the farther apart two linked alleles are on a chromosome, the more often the chromosome will cross over between them. This finding led to recombination mapping—mapping of linkage groups with each map unit being equal to 1 percent recombination. For example, if two linked genes, *A* and *B*, recombine with a frequency of 15 percent, and *B* and *C* recombine with a frequency of 9 percent, and *A* and *C* recombine with a frequency of 24 percent, what is the sequence and the distance between them?

The sequence and the distance of *A-B-C* is

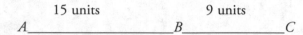

If the recombination frequency between *A* and *C* had been 6 percent instead of 24 percent, the sequence and distance of *A-B-C* would instead be

SEX-LINKED TRAITS

We already know that humans contain 23 pairs of chromosomes. Twenty-two of the pairs of chromosomes are called **autosomes**. They code for many different traits. The other pair contains the **sex chromosomes**. This pair determines the sex of an individual. A female has two X chromosomes. A male has one X and one Y chromosome—an X from his mother and a Y from his father. Some traits, such as **color blindness** and **hemophilia**, are carried on sex chromosomes. These are called **sex-linked traits**. Most sex-linked traits are found on the X chromosome and are more properly referred to as "X-linked."

Since males have one X and one Y chromosome, what happens if a male has an X-chromosome with the color blindness allele? Unfortunately, he'll express the sex-linked trait, even if it is recessive. Why? Because his one and only X chromosome is color-blind X. He doesn't have another X to mask the effect of the color-blind X. However, if a female has only one color-blind X-chromosome, she won't express a recessive sex-linked trait. For her to express the trait, she has to inherit two color-blind X-chromosomes. A female with one color-blind X is called a **carrier**. Although she does not exhibit the trait, she can still pass it on to her children.

You can also use the Punnett square to figure out the results of sex-linked traits. Here's a classic example: a male who has normal color vision and a woman who is a carrier for color blindness have children. How many of the children will be color-blind?

> Regarding this topic, we are following what the AP currently teaches in terms of biological sex (not gender), which is defined by sex chromosomes, with female being XX and male being XY.

> Both red/green color blindness and hemophilia are X-linked recessive traits. Other X-linked traits are dominant, meaning only one copy of the mutant allele is required for expression of the trait in females.

AP Biology Premium Prep

You will not need to know other X-linked conditions (like Rett Syndrome or hemophilia) on test day, but if you are curious, you can find a short addendum about such diseases in your online Student Tools.

To figure out the answer, let's set up a Punnett square.

$$\bar{X} = \text{color-blind X}$$

Notice that we placed a bar above any color-blind X to indicate the presence of an allele for color blindness. And now for the results. The couple could have a son who is color-blind, a son who sees color normally, a daughter who is a carrier, or a daughter who sees color normally.

Barr Bodies

A look at the cell nucleus of normal females will reveal a dark-staining body known as a **Barr body**. A Barr body is an X chromosome that is condensed and visible. In every female cell, one X chromosome is activated and the other X chromosome is deactivated during embryonic development. Surprisingly, the X chromosome destined to be inactivated is randomly chosen in each cell. Therefore, in every tissue in the adult female, one X chromosome remains condensed and inactive. However, this X chromosome is replicated and passed on to a daughter cell. X-inactivation is the reason it is okay that females have two X chromosomes and males have only one. After X-inactivation, it is like everyone has one copy.

Other Inheritance Patterns

Not all patterns of inheritance obey the principles of Mendelian genetics. In fact, many traits we observe are due to a combined expression of alleles. Here are a couple of examples of non-Mendelian forms of inheritance:

- **Incomplete dominance (blending inheritance):** In some cases, the traits will blend. For example, if you cross a white snapdragon plant (genotype *WW*) with a red snapdragon plant (*RR*), the resulting progeny will be pink (*RW*). In other words, neither color is dominant over the other.
- **Codominance:** Sometimes you'll see an equal expression of both alleles. For example, an individual can have an AB blood type. In this case, each allele is equally expressed. That is, both the A allele and the B allele are expressed (I^AI^B). That's why the person is said to have AB blood. The expression of one allele does not prevent the expression of the other.
- **Polygenic inheritance:** In some cases, a trait results from the interaction of *many* genes. Each gene will have a small effect on a particular trait. Height, skin color, and weight are all examples of polygenic traits.

- **Non-nuclear inheritance**: Apart from the genetic material held in the nucleus, there is also genetic material in the mitochondria. The mitochondria are always provided by the egg during sexual reproduction, so mitochondrial inheritance is always through the maternal line, not the male line. In plants, the mitochondria are provided by the ovule and are maternally inherited. Chloroplasts also contain DNA, in the form of one circular double-stranded molecule. Both chloroplasts and mitochondria assort randomly into daughter cells and gametes.

Pedigrees

One way to study genetic inheritance is by looking at a special family tree called a **pedigree**. A pedigree shows which family members have a particular trait and it can help determine if a trait is recessive or dominant and if it is sex-linked. Traits that skip generations are usually recessive. Traits that appear more in one sex than the other are usually sex-linked. In a pedigree chart, the males are squares and the females are circles. Matings are drawn with horizontal lines and the offspring are shown below the mated couple.

ENVIRONMENTAL EFFECTS ON TRAITS

Changes in genotypes can result in changes in phenotype, but environmental factors also influence many traits, both directly and indirectly. In addition, an organism's adaptation to the local environment reflects a flexible response of its genome. For example, gene expression and protein function can be affected by temperature. **Phenotypic plasticity** occurs if two individuals with the same genotype have different phenotypes because they are in different environments.

Here are some examples of how the environment can influence traits both directly and indirectly:

Height and Weight in Humans
- Controlled by both genetic and environmental factors.
- **Examples:** nutrition, disease, physical activity, medication use

Soil pH Affects Flower Color
- Soil pH affects nutrient accessibility, which can change flower color.
- In acidic soil (pH < 5.5), aluminum is available to the roots of hydrangeas, resulting in blue flowers.
- When no aluminum is available, flowers are white.

Seasonal Fur Color in Arctic Animals
- Arctic hare are brown or gray in the summer, when their habitat is full of greens and browns.
- Arctic hare are white in winter because everything is covered in snow.
- Color change is linked to photoperiod (or how much light is received during the day).

Sex Determination in Reptiles
- Temperatures experienced during embryonic/larval development determine the sex of offspring.
- In lizards, male embryos develop at intermediate temperatures and female embryos are generated at both extremes.

AN OVERVIEW OF MEIOSIS

Meiosis is the production of gametes. Since sexually reproducing organisms need only haploid cells for reproduction, meiosis is limited to sex cells in special sex organs called **gonads**. In males, the gonads are the **testes**, while in females they are the **ovaries**. The special cells in these organs—also known as **germ cells**—produce haploid cells (n), which then combine to restore the diploid ($2n$) number during fertilization.

$$\text{female gamete } (n) + \text{male gamete } (n) = \text{zygote } (2n)$$

When it comes to genetic variation, meiosis is a big plus. Variation, in fact, is the driving force of evolution. The more variation there is in a population, the more likely it is that some members of the population will survive extreme changes in the environment. Meiosis is far more likely to produce these sorts of variations than is mitosis, which therefore confers selective advantage on sexually reproducing organisms. We'll come back to this theme in Chapter 10.

A Closer Look at Meiosis

Meiosis actually involves two rounds of cell division: **meiosis I** and **meiosis II**.

Before meiosis begins, the diploid cell goes through interphase. Just as in mitosis, double-stranded chromosomes are formed during S phase.

Meiosis I

Meiosis I consists of four stages: prophase I, metaphase I, anaphase I, and telophase I. Meiosis I ensures that each gamete receives a haploid ($1n$) set of chromosomes.

Prophase I

Prophase I is a little more complicated than regular prophase. As in mitosis, the nuclear membrane disappears, the chromosomes become visible, and the centrioles move to opposite poles of the nucleus. But that's where the similarity ends.

The major difference involves the movement of the chromosomes. In meiosis, the chromosomes line up side-by-side with their counterparts (homologs). This event is known as **synapsis**.

Synapsis involves two sets of chromosomes that come together to form a **tetrad** (a **bivalent**). A tetrad consists of four chromatids. Synapsis is followed by **crossing-over**, the exchange of segments between homologous chromosomes.

What's unique in prophase I is that pieces of chromosomes are exchanged between homologous partners. This is one of the ways organisms produce genetic variation. By the end of prophase I, chromosomes will have exchanged regions containing several **alleles**, or different forms of the same gene. This means that each chromosome will have a different mix of alleles than it began with. Remember, the two homologous chromosomes are versions that came from the person's mother and father. So, a chromosome that has experienced crossing-over will now be a mix of alleles from the mother and father versions.

Metaphase I

As in mitosis, the chromosome pairs—now called tetrads—line up at the metaphase plate. By contrast, you'll recall that in regular metaphase, the chromosomes line up individually. One important concept to note is that the alignment during metaphase is random, so the copy of each chromosome that ends up in a daughter cell is random. Therefore, the gamete will be created with a mixture of genetic information from the person's father and mother. This means that the offspring created would be a combination of all four grandparents.

Metaphase Plate

Anaphase I

During anaphase I, each pair of chromatids within a tetrad moves to opposite poles. Notice that the chromatids do not separate at the centromere. The homologs separate with their centromeres intact.

The chromosomes now move to their respective poles.

Telophase I
During telophase I, the nuclear membrane forms around each set of chromosomes.

Finally, the cells undergo cytokinesis, leaving us with two daughter cells. Notice that at this point the nucleus contains the haploid number of chromosomes, but each chromosome is a duplicated chromosome, which consists of two chromatids.

Meiosis II

The purpose of the second meiotic division is to separate sister chromatids, and the process is virtually identical to mitosis. Let's run through the steps of meiosis II.

During prophase II, chromosomes once again condense and become visible. In metaphase II, chromosomes move toward the metaphase plate. This time they line up single file, not as pairs. During anaphase II, chromatids of each chromosome split at the centromere, and each chromatid is pulled to opposite ends of the cell. At telophase II, a nuclear membrane forms around each set of chromosomes and a total of *four* haploid cells are produced.

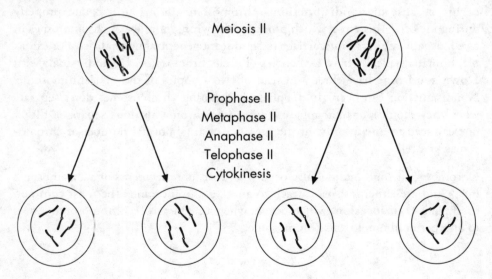

8. Heredity | 197

Gametogenesis

Meiosis is also known as **gametogenesis.** If sperm cells are produced, then meiosis is called **spermatogenesis**. During spermatogenesis, four sperm cells are produced for each diploid cell. If an egg cell or an ovum is produced, this process is called **oogenesis**.

Oogenesis is a little different from spermatogenesis. Oogenesis produces only one ovum, not four. The other three cells, called **polar bodies,** get only a tiny amount of cytoplasm and eventually degenerate. Why does oogenesis produce only one ovum? Because the female wants to conserve as much cytoplasm as possible for the surviving gamete, the **ovum**.

Here's a summary of the major differences between mitosis and meiosis.

MITOSIS	MEIOSIS
• Occurs in somatic (body) cells	• Occurs in germ (sex) cells
• Produces identical cells	• Produces gametes
• Diploid cell → diploid cells	• Diploid cell → haploid cells
• 1 cell becomes 2 cells	• 1 cell becomes 4 cells
• Number of divisions: 1	• Number of divisions: 2

Meiotic Errors

Down syndrome is also known as trisomy 21. Two other conditions you should be familiar with are Turner Syndrome (45XO), which occurs when a female is missing part of or a whole X chromosome, and Klinefelter's Syndrome (47XXY), which occurs when a male has an extra X chromosome.

Sometimes, a set of chromosomes has an extra or a missing chromosome. This occurs because of **nondisjunction**—chromosomes failed to separate properly during meiosis. This error, which produces the wrong number of chromosomes in a cell, usually results in miscarriage or significant genetic abnormalities. For example, humans typically have two copies of each chromosome, but individuals with **Down syndrome** have three—instead of two—copies of the 21st chromosome. Nondisjunction can occur in anaphase I (meaning chromosomes don't separate when they should), or in anaphase II (meaning chromatids don't separate). Either one can lead to **aneuploidy**, or the presence of an abnormal number of chromosomes in a cell.

Chromosomal abnormalities also occur if one or more segments of a chromosome break and are either lost or reattach to another chromosome. The most common example is **translocation**, which occurs when a segment of a chromosome moves to another nonhomologous chromosome.

Here's an example of a translocation.

Translocations can occur when recombination (crossing-over) occurs incorrectly. Fortunately, in most cases, damaged DNA can usually be repaired by special repair enzymes.

Many ethical, social, and medical issues surround human genetic disorders. Individuals who are carriers for genetic diseases are at increased risk of having a child with a genetic disease. A prenatal diagnosis can identify some genetic diseases before birth, and parents can then prepare to have a child with special needs or have the option of terminating the pregnancy. In addition, individuals who carry genetic diseases sometimes use an egg or sperm donor instead of using their own gametes.

Civic issues are also important to consider. Genetic discrimination occurs when people are treated unfairly because of actual or perceived differences in their genetic information that may cause or increase the risk of developing a disorder or disease. For example, a health insurer might refuse to give coverage to an individual who has a genetic difference that raises their odds of getting cancer. Many G7 countries protect genetic test information to help eliminate genetic discrimination and safeguard individuals' genetic privacy.

CHAPTER 8 KEY TERMS

diploid cell
haploid cell
homologous chromosomes
sex cells (or gametes)
Gregor Mendel
trait
genes
locus
alleles
homozygous
heterozygous
phenotype
genotype
dominant
recessive
parent, or P generation
filial, or F_1 generation
F_2 generation
Law of Dominance
Law of Segregation
Law of Independent Assortment
monohybrid cross
Punnett square
dihybrid cross
Product Rule
Sum Rule
test cross
linked genes
recombinants
percentage of recombination
    (recombination frequency)
map units (centimorgans)
autosomes

sex chromosomes
color blindness
hemophilia
sex-linked traits
carrier
Barr body
incomplete dominance
    (blending inheritance)
codominance
polygenetic inheritance
non-nuclear inheritance
pedigree
phenotypic plasticity
meiosis
gonads
testes
ovaries
germ cells
meiosis I
meiosis II
synapsis
tetrad (bivalent)
crossing-over (recombination)
gametogenesis
spermatogenesis
oogenesis
polar bodies
ovum
nondisjunction
Down syndrome
aneuploidy
translocation

AP Biology Premium Prep

Chapter 8 Drill

Answers and explanations can be found in Chapter 15.

---

1 ☐ Mark for Review

Two expecting parents wish to determine all possible blood types of their unborn child. If both parents have an AB blood type, which of the following blood types will their child have the greatest chance of possessing?

(A) A

(B) B

(C) AB

(D) O

---

2 ☐ Mark for Review

In pea plants, the trait for round seeds (R) is dominant to the trait for wrinkled seeds (r). The trait for purple flowers (P) is dominant to white flowers (p). A researcher crosses two pea plants and finds the probability of an offspring having round seeds is 50% and the probability of a plant having purple flowers is 75%. Which of the following combinations most likely represents the genotypes of the parents?

(A) $RrPp \times RrPp$

(B) $RrPp \times rrpp$

(C) $RrPp \times rrPp$

(D) $RRPp \times rrPp$

---

3 ☐ Mark for Review

In a pedigree, females are typically denoted with circles and males with squares. Individuals affected by the condition are typically shaded, but this does not indicate whether the condition is dominant or recessive. Which of the following would you expect to find in a pedigree when a son has an X-linked recessive disorder?

(A) All squares representing the offspring of a shaded circle will be shaded.

(B) Only circles in the pedigree will be shaded.

(C) All squares and circles in the first generation will be shaded.

(D) Only squares in the pedigree will be shaded.

---

4 ☐ Mark for Review

Two genes, W and Z, are on two different chromosomes. The probability of gene W being passed on to a gamete is 50%, while the probability of gene Z being passed on to a gamete is 25%. What is the probability that both genes W and Z will be passed on to the same gamete?

(A) 0.5 + 0.25

(B) 0.5 / 0.25

(C) 0.25 / 0.5

(D) 0.5 × 0.25

---

8. Heredity

AP Biology Premium Prep

---

5 ☐ Mark for Review

Crossover frequencies can be used to map the relative location of genes on a chromosome. The data below shows the crossover frequencies of three genes on the same chromosome.

Genes	Crossover Frequency
7%	X and Z
17%	Y and W
32%	X and Y
47%	Y and X
52%	Y and Z

Which of the following represents the relative locations of these three genes on the chromosome?

(A) WZYX

(B) WXYZ

(C) YWXZ

(D) YWZX

---

6 ☐ Mark for Review

In humans, the presence of dimples is controlled by a single gene with two alleles. When a homozygous person with dimples reproduces with a homozygous person without dimples, only offspring with dimples are born. If two people with the same genotype of the offspring reproduce with each other, what percent of their offspring will most likely have dimples?

(A) 0%

(B) 25%

(C) 75%

(D) 100%

---

7 ☐ Mark for Review

The table below shows the number of chromosomes present in autosomal cells of five different species that reproduce by sexual reproduction.

Species	# of Chromosomes
Human	46
House cat	38
Gorilla	48
Dolphin	44
Cheetah	38

Which of the following statements accurately describes the number of chromosomes in the gametes of each of these species?

(A) A gorilla's gametes will have 24 chromosomes, which is more than any other species in the table.

(B) A house cat and a cheetah will both have 38 chromosomes in their gametes, which is less than any other species in the table.

(C) A human's gametes will have 46 chromosomes, which is more than any other species in the table.

(D) A cheetah and house cat will be able to reproduce because they have the same number of chromosomes.

---

202 | For more free content, visit PrincetonReview.com

8. Mark for Review

Marfan syndrome is a genetic disorder that results in long fingers and toes, a curved spine, headaches, flat feet, and a fast heart rate in affected individuals. The pedigree below indicates the incidence of Marfan syndrome in a family. A genetic counselor advises the family that if the individuals in P1 reproduce again, the fourth child will have a 50% chance of having Marfan syndrome. Based on this information, which of the following is the most likely type of inheritance for Marfan syndrome?

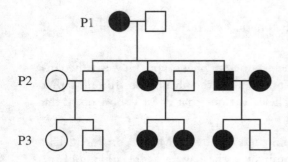

(A) X-linked dominant

(B) X-linked recessive

(C) Autosomal dominant

(D) Autosomal recessive

9. Mark for Review

A scientist performs an experiment in which heat lamps are placed near a snake in the process of shedding its skin. The new scales that appear on the snake are a different color than the original ones. Which of the following conclusions is best supported by this experiment?

(A) Snakes who can change the color of their scales are a more evolved species.

(B) The snake's genes are flexible and can change in response to stimuli in the environment.

(C) Heat energy can adjust the genome of snakes and cause new traits to appear.

(D) The snakes can choose to change their phenotypes in order to shed their skins more efficiently.

10. Mark for Review

A couple has a child with Triple X syndrome, and she has a genotype of XXX. Which of the following could be the error that caused the syndrome to occur in this child?

(A) Trisomy when the egg gamete was formed

(B) Crossing-over between the X and Y chromosomes in the sperm gamete

(C) Nondisjunction when the egg gamete was formed

(D) Mitotic errors after the egg gamete was fertilized by a sperm gamete

Chapter 8 Summary

- Genetics is the study of heredity. The foundation of genetics is that traits are produced by genes, which are found on chromosomes.
- Diploid organisms carry two copies of every gene, one on each of two homologous chromosomes, one that came from their mother and one that came from their father.
- When two alleles are the same, the organism is homozygous for that trait. When the two alleles are different, the organism is heterozygous for the trait.
- Genes are described as the genotype of the organism. The physical expression of the traits, or what you see in the organism, is the phenotype.
- Homozygous individuals have only one allele that will determine the phenotype. However, heterozygous individuals have two different alleles that could contribute to the phenotype. The allele that "wins out" and directs the phenotype is the dominant allele. The one that doesn't show up in the phenotype is the recessive allele.
- Simple genetics with dominants and recessives was illuminated by Gregor Mendel. In Mendelian genetics, there are three important laws: the Law of Dominance, the Law of Segregation, and the Law of Independent Assortment.
- If two genes are on the same chromosome and close together, they are linked. Completely linked alleles on the same chromosome are locked together and travel together during gamete formation. Recombination frequency can be used to measure how far apart two genes are on a chromosome.
- Humans have 23 pairs of homologous chromosomes for a total of 46. Of these, 22 pairs are autosomes and 2 are designated as sex chromosomes:
  - Females have two X chromosomes; males have one X and one Y chromosome.
  - Because females have two X chromosomes, one is inactivated in each cell and condenses into a Barr body.
- Non-Mendelian genetics refers to a situation in which traits do not follow the Mendelian laws. Examples include:
  - incomplete dominance
  - codominance
  - polygenic inheritance

- Meiosis produces four genetically distinct haploid gametes (sperm or egg cells). Meiosis involves two rounds of cell division: meiosis I and meiosis II.
  - During meiosis I, the homologous chromosome pairs separate, but before they can, they undergo crossing-over and swap some DNA.
  - During meiosis II, sister chromatids separate.
  - Mutations, such as nondisjunction events or whole chromosome translocations, can lead to genetic diseases.

Chapter 9
Gene Expression and Regulation

DNA: THE BLUEPRINT OF LIFE

All living things possess an astonishing degree of organization. From the simplest single-celled organism to the largest mammal, millions of reactions and events must be coordinated precisely for life to exist. This coordination is directed by DNA, which is the hereditary blueprint of the cell. Some organisms rely on RNA to transfer information from one generation to the next.

DNA is made up of repeated subunits of **nucleotides**. Each nucleotide has a **five-carbon sugar,** a **phosphate**, and a **nitrogenous base**. Take a look at the nucleotide below.

The name of the pentagon-shaped sugar in DNA is **deoxyribose**. Hence, the name *deoxyribo*nucleic acid. Notice that the sugar is linked to two things: a phosphate and a nitrogenous base. A nucleotide can have one of four different nitrogenous bases:

- **adenine**—a purine (double-ringed nitrogenous base)
- **guanine**—a purine (double-ringed nitrogenous base)
- **cytosine**—a pyrimidine (single-ringed nitrogenous base)
- **thymine**—a pyrimidine (single-ringed nitrogenous base)

Any of these four bases can attach to the sugar. As we'll soon see, this is extremely important when it comes to the message of the genetic code in DNA. Prokaryotes and eukaryotes can also contain **plasmids**, which are small, double-stranded, circular DNA molecules.

The nucleotides can link up in a long chain to form a single strand of DNA. Here's a small section of a DNA strand.

The nucleotides themselves are linked together by **phosphodiester bonds** between the sugars and the phosphates. This is called the sugar-phosphate backbone of DNA and it serves as a scaffold for the bases.

Two DNA Strands

Each DNA molecule consists of two strands that wrap around each other to form a long, twisted ladder called a **double helix**. The structure of DNA was brilliantly deduced in 1953 by three scientists: **Watson, Crick, and Franklin.**

Now let's look at the way two DNA strands get together. Again, think of DNA as a ladder. The sides of the ladder consist of alternating sugar and phosphate groups, while the rungs of the ladder consist of pairs of nitrogenous bases.

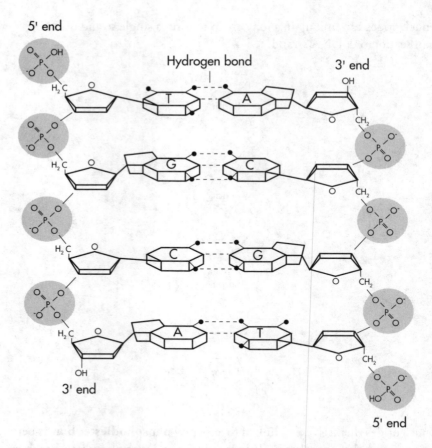

The nitrogenous bases pair up in a particular way. Adenine in one strand always binds to thymine (A–T or T–A) in the other strand. Similarly, guanine always binds to cytosine (G–C or C–G). This predictable matching of the bases is known as **base pairing**.

The two strands are said to be **complementary**. This means that if you know the sequence of bases in one strand, you'll know the sequence of bases in the other strand. For example, if the base sequence in one DNA strand is A–T–C, the base sequence in the complementary strand will be T–A–G.

The two DNA strands run in opposite directions. You'll notice from the figure above that each DNA strand has a 5' end and a 3' end, so-called for the carbon that ends the strand (i.e., the fifth carbon in the sugar ring is at the 5' end, while the third carbon is at the 3' end). The 5' end has a phosphate group, and the 3' end has an OH, or "hydroxyl," group. The 5' end of one strand is always opposite to the 3' end of the other strand. The strands are therefore said to be **antiparallel**.

The DNA strands are linked by **hydrogen bonds**. Two hydrogen bonds hold adenine and thymine together, and three hydrogen bonds hold cytosine and guanine together.

Before we go any further, let's review the base pairing in DNA:

- Adenine pairs up with thymine (A–T or T–A) by forming two hydrogen bonds.
- Cytosine pairs up with guanine (G–C or C–G) by forming three hydrogen bonds.

GENOME STRUCTURE

The order of the four base pairs in a DNA strand is the genetic code. Like a special alphabet in our cells, these four nucleotides spell out thousands of recipes. Each recipe is called a **gene**. The human genome has around 20,000 genes.

The recipes of the genes are spread out among the millions of nucleotides of DNA, and all of the DNA for a species is called its **genome**. Each separate chunk of DNA in a genome is called a **chromosome**.

Prokaryotes have one circular chromosome, and eukaryotes have linear chromosomes. In eukaryotes, the DNA is further structured, probably because linear chromosomes are more likely to get tangled. To keep it organized, the DNA is wrapped around proteins called **histones**, and then the histones are bunched together in groups called a **nucleosome**. Not all DNA is wound up equally because it must be unwound in order to be read. How tightly DNA is packaged depends on the section of DNA and also what is going on in the cell at that time. When the genetic material is in a loose form in the nucleus, it is called **euchromatin,** and its genes are active, or available for transcription. When the genetic material is fully condensed into coils, it is called **heterochromatin,** and its genes are generally inactive. Situated in the nucleus, chromosomes contain the recipe for all the processes necessary for life, including passing themselves and their information on to future generations. In this chapter, we'll look at precisely how they accomplish this.

> Humans have 23 different chromosomes and they have two copies of each, so each human cell has 46 linear chromosome segments in the nucleus.

DNA REPLICATION

We said in the beginning of this chapter that DNA is the hereditary blueprint of the cell. By directing the manufacture of proteins, DNA serves as the cell's blueprint. But how is DNA inherited? For the information in DNA to be passed on, it must first be copied. This copying of DNA is known as **DNA replication**.

Because the DNA molecule is twisted over on itself, the first step in replication is to unwind the double helix by breaking the hydrogen bonds. This is accomplished by an enzyme called **helicase**. The exposed DNA strands now form a *y*-shaped **replication fork.**

Now each strand can serve as a template for the synthesis of another strand. DNA replication begins at specific sites called **origins of replication**. Because the DNA helix twists and rotates during DNA replication, another class of enzymes, called

> Neither DNA strand is directly copied; however, by building a new partner strand for each of them, a two-stranded molecule is created that is identical to the original.

DNA **topoisomerases**, cuts and rejoins the helix to prevent tangling. The enzyme that performs the actual addition of nucleotides to the freshly built strand is **DNA polymerase**. But DNA polymerase, oddly enough, can add nucleotides only to the 3' end of an existing strand. Therefore, to start off replication, an enzyme called **RNA primase** adds a short strand of RNA nucleotides called an **RNA primer**. After replication, the primer is degraded by enzymes and replaced with DNA so that the final strand contains only DNA.

During DNA replication, one DNA strand is called the **leading strand**, and it is made continuously. That is, the nucleotides are steadily added one after the other by DNA polymerase. The other strand—the **lagging strand**—is made discontinuously. Unlike the leading strand, the lagging strand is made in pieces of nucleotides known as **Okazaki fragments**.

Why is the lagging strand made in small pieces? Nucleotides are added only in the 5' to 3' direction since nucleotides can be added only to the 3' end of the growing chain. However, when the double-helix is "unzipped," one of the two strands is oriented in the opposite direction—3' to 5'. Because DNA polymerase doesn't work in this direction, the strand needs to be built in pieces. You can see in the figure below that the leading strand is being created toward the opening of the helix and the helix continually opens ahead of it to accommodate it. The lagging strand is built in the opposite direction of the way the helix is opening, so it can build only until it hits a previously built stretch. Once the helix unwinds a bit more, it can build another Okazaki fragment, and so on. These fragments are eventually linked together by the enzyme **DNA ligase** to produce a continuous strand. Finally, hydrogen bonds form between the new base pairs, leaving two identical copies of the original DNA molecule.

When DNA is replicated, we don't end up with two entirely new molecules. Each new molecule has half of the original molecule. Because DNA replicates in a way that conserves half of the original molecule in each of the two new ones, it is said to be **semiconservative**.

An interesting problem is that a few bases at the very end of a DNA molecule cannot be replicated because DNA polymerase needs space to bind template DNA. This means that every time replication occurs, the chromosome loses a few base pairs at the end. The genome has compensated for this over time, by putting bits of unimportant (or at least less important) DNA at the ends of a molecule. These ends are called **telomeres**. They get shorter and shorter over time.

Many enzymes and proteins are involved in DNA replication. The ones you'll need to know for the AP Biology Exam are DNA helicase, DNA polymerase, DNA ligase, topoisomerase, and RNA primase:

- Helicase unwinds our double helix into two strands.
- DNA polymerase adds nucleotides to an existing strand.
- Ligase brings together the Okazaki fragments.
- Topoisomerase cuts and rejoins the helix.
- RNA primase catalyzes the synthesis of RNA primers.

> If you are interested in some of the history behind the early DNA experiments, we have included a short addendum in your online Student Tools. It will not, however, be tested.

Central Dogma

DNA's main role is directing the manufacture of molecules that actually do the work in the body. DNA is just the recipe book, not the chef or the meal. When a DNA recipe is used, scientists say that that DNA is expressed.

The first step of DNA expression is to turn it into RNA. The RNA is then sent out into the cell and often gets turned into a protein. Proteins are the biggest group of "worker-molecules," and most expressed DNA turns into proteins. These proteins, in turn, regulate almost everything that occurs in the cell.

The process of making an RNA from DNA is called **transcription,** and the process of making a protein from an RNA is called **translation**. In eukaryotic cells, transcription takes place in the nucleus (where the DNA is kept), and translation takes place in the cytoplasm.

Obviously, in prokaryotes, both occur in the cytoplasm because prokaryotes don't contain a nucleus; this allows transcription and translation to occur at the same time.

The flow of genetic information, the **Central Dogma of Biology**, is therefore:

$$\text{DNA} \xrightarrow[\text{in the nucleus}]{\text{transcription}} \text{RNA} \xrightarrow[\text{in the cytoplasm}]{\text{translation}} \text{Proteins}$$

RNA

Before we discuss transcription of RNA, let's talk about its structure. Although RNA is also made up of nucleotides, it differs from DNA in three ways.

1. RNA is single-stranded, not double-stranded.
2. The five-carbon sugar in RNA is **ribose** instead of deoxyribose.
3. The RNA nitrogenous bases are adenine, guanine, cytosine, and a different base called **uracil**. Uracil replaces thymine as adenine's partner.

> **Word Connection**
> A dogma is a core belief or set of ideas. Since this flow of information from DNA to RNA to protein is so important for understanding the workings of living things, it has been dubbed the Central Dogma of Biology.

Here's a table to compare DNA and RNA. Keep these differences in mind—the testing board loves to test you on them.

DIFFERENCES BETWEEN DNA AND RNA		
	DNA (double-stranded)	**RNA (single-stranded)**
Sugar	deoxyribose	ribose
Bases	adenine	adenine
	guanine	guanine
	cytosine	cytosine
	thymine	uracil

Think Critically

Can you think of any exceptions to the Central Dogma? Do all genes get turned into RNA and then into proteins?

Hint: rRNA, tRNA, and RNAi never get turned into protein; they stay as RNA.

There are three main types of RNA: messenger RNA (mRNA), ribosomal RNA (rRNA), and transfer RNA (tRNA). All three types of RNA are key players in the synthesis of proteins:

- **Messenger RNA (mRNA)** is a temporary RNA version of a DNA recipe that gets sent to the ribosome.
- **Ribosomal RNA (rRNA)**, which is produced in the nucleolus, makes up part of the ribosomes. You'll recall from our discussion of the cell in Chapter 5 that the ribosomes are the sites of protein synthesis. We'll see how they function a little later on.
- **Transfer RNA (tRNA)** shuttles amino acids to the ribosomes. It is responsible for bringing the appropriate amino acids into place at the appropriate time by matching anticodons to codons. It does this by reading the message carried by the mRNA.

There is also another class of RNA, called interfering RNAs, or **RNAi.** These are small snippets of RNA that are naturally made in the body or intentionally created by humans. These interfering RNAs, called siRNA and miRNA, can bind to specific sequences of RNA and mark them for destruction. This will be discussed more later in this chapter. Now that we know about the different types of RNA, let's see how they direct the synthesis of proteins.

Transcription

Transcription involves making an RNA copy of a bit of DNA code. The initial steps in transcription are similar to the initial steps in DNA replication. The obvious difference is that, whereas in replication we end up with a complete copy of the cell's DNA, in transcription we end up with only a tiny specific section copied into an mRNA. This is because only the bit of DNA that needs to be expressed will be transcribed. If you wanted to make a cake, but your cookbook couldn't leave your vault (aka the nucleus), you wouldn't copy the entire cookbook! You would copy only the recipe for the cake.

Since each recipe is a gene, transcription occurs as-needed on a gene-by-gene basis.

The exception to this is prokaryotes because they will transcribe a recipe that can be used to make several proteins. This is called a **polycistronic transcript**. Eukaryotes tend to have one gene that gets transcribed to one mRNA and translated into one protein. These transcripts are **monocistronic.**

Transcription involves three phases: initiation, elongation, and termination. As in DNA replication, the first initiation step in transcription is to unwind and unzip the DNA strands using helicase. Transcription begins at special sequences of the DNA strand called **promoters**. You can think of a promoter as a docking site or a runway. We will talk about how promoters are involved in regulating transcription later in the chapter.

Because RNA is single-stranded, we have to copy only *one* of the two DNA strands. The strand that serves as the template is known as the **antisense strand,** the **noncoding strand, minus-strand**, or the **template strand**. The other strand that lies dormant is the **sense strand,** or the **coding strand**.

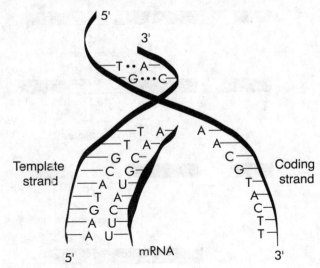

Just as DNA polymerase builds DNA, **RNA polymerase** builds RNA, and just like DNA polymerase, RNA polymerase adds nucleotides only to the 3' side, therefore building a new molecule from 5' to 3'. This means that RNA polymerase must bind to the 3' end of the template strand first (which would be the 5' end of future mRNA).

RNA polymerase doesn't need a primer, so it can just start transcribing the DNA right off the bat. The promoter region is considered to be "upstream" of the actual coding part of the gene. This way the polymerase can get set up before the bases it needs to transcribe, like a staging area before an official parade starting point. The official starting point is called the **start site**.

When transcription begins, RNA polymerase travels along and builds an RNA that is complementary to the template strand of DNA. It is just like DNA replication except when DNA has an adenine, the RNA can't add a thymine since RNA doesn't have thymine. Instead, the enzyme adds a uracil.

Once RNA polymerase finishes adding on nucleotides and reaches the termination sequence, it separates from the DNA template, completing the process of transcription. The new RNA is now a transcript or a copy of the sequence of nucleotides based on the DNA strand. Note that the freshly synthesized RNA is complementary to the template strand, but it is identical to the coding strand (with the substitution of uracil for thymine).

RNA Processing

In prokaryotes, the mRNA is now complete, but in eukaryotes the RNA must be processed before it can leave the nucleus.

In eukaryotes, the freshly transcribed RNA is called an hnRNA (heterogeneous nuclear RNA) and it contains both coding regions and noncoding regions. The regions that express the code that will be turned into protein are **exons**. The noncoding regions in the mRNA are **introns**.

The introns—the intervening sequences—must be removed before the mRNA leaves the nucleus. This process, called **splicing**, is accomplished by an RNA-protein complex called a **spliceosome**. This process produces a final mRNA that is shorter than the transcribed RNA. The way that a transcript is spliced can vary, and alternative splicing variants will result with different exons included.

In addition to splicing, a **poly(A) tail** is added to the 3' end and a **5' GTP cap** is added to the 5' end.

TRANSLATION

Translation is the process of turning an mRNA into a protein that occurs on ribosomes in cytoplasm and on the rough endoplasmic reticulum. Remember, each protein is made of amino acids. The order of the mRNA nucleotides will be read in the ribosome in groups of three. A group of three nucleotides is called a **codon**. Each codon corresponds to a particular amino acid. The genetic code is redundant, meaning that certain amino acids are specified by more than one codon. Nearly all living organisms use the same genetic code, which is further evidence for a common ancestor.

The mRNA attaches to the ribosome to initiate translation and "waits" for the appropriate amino acids to come to the ribosome. That's where tRNA comes in. A tRNA molecule has a unique three-dimensional structure that resembles a four-leaf clover:

One end of the tRNA carries an amino acid. The other end, called an **anticodon**, has three nitrogenous bases that can complementarily base pair with the codon in the mRNA. Usually, the normal rules of base pairing are set in stone, but tRNA anticodons can be a bit flexible when they bind with a codon on an mRNA, especially the third nucleotide in an anticodon. The third position is said to experience **wobble pairing**. Things that don't normally bind will pair up, like guanine and uracil.

Transfer RNAs are the "go-betweens" in protein synthesis. Each tRNA becomes charged and enzymatically attaches to an amino acid in the cell's cytoplasm and "shuttles" it to the ribosome. The charging enzymes involved in forming the bond between each amino acid and tRNA require ATP.

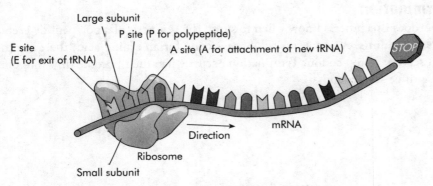

Note: Each site, E, P, or A, is a groove for the tRNA to fit in.

Translation also involves three phases: **initiation**, **elongation**, and **termination**. Initiation begins when a ribosome attaches to the mRNA.

What does the ribosome do? It helps the process along by holding everything in place while the tRNAs assist in assembling polypeptides.

Initiation

Ribosomes contain three binding sites: an **A site**, a **P site**, and an **E site**. The mRNA will shuffle through from A to P to E. As the mRNA codons are read, the polypeptide will be built. In all organisms, the **start codon** for the initiation of protein synthesis is A–U–G, which codes for the amino acid methionine. Proteins can have AUGs in the rest of the protein that code for methionine as well, but the special first AUG of an mRNA is the one that will kick off translation. The tRNA with the complementary anticodon, U–A–C, is methionine's personal shuttle; when the AUG is read on the mRNA, methionine is delivered to the ribosome.

Elongation

Addition of amino acids is called elongation. Remember that the mRNA contains many codons, or "triplets," of nucleotide bases. As each amino acid is brought to the mRNA by tRNA molecules, it is linked to its neighboring amino acid by a peptide bond. When many amino acids link up, a polypeptide is formed.

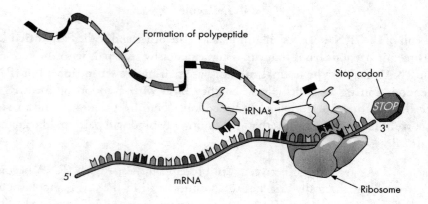

Termination

How does this process know when to stop? The synthesis of a polypeptide is ended by **stop codons**. A codon doesn't always code for an amino acid; there are three that serve as a stop codon. Termination occurs when the ribosome runs into one of these three stop codons.

How about a little review?

- In transcription, mRNA is created from a particular gene segment of DNA.
- In eukaryotes, the mRNA is "processed" by having its introns, or noncoding sequences, removed. A 5' cap and a 3' tail are also added.
- Now, ready to be translated, mRNA proceeds to the ribosome.
- Free-floating amino acids are picked up by tRNA and shuttled over to the ribosome, where mRNA awaits.
- In translation, the anticodon of a tRNA molecule carrying the appropriate amino acid base pairs with the codon on the mRNA.
- As new tRNA molecules match up to new codons, the ribosome holds them in place, allowing peptide bonds to form between the amino acids.
- The newly formed polypeptide grows until a stop codon is reached.
- The polypeptide or protein folds up and is released into the cell.

At the Same Time!
In prokaryotes, transcription and translation can be occurring at the same time. This is because prokaryotes lack a nucleus and their transcription and translation are both occurring in the same place, the cytoplasm.

1st Position (5' End)	2nd Position				3rd Position (3' End)
	U	C	A	G	
U	Phe	Ser	Tyr	Cys	U
	Phe	Ser	Tyr	Cys	C
	Leu	Ser	**Stop**	**Stop**	A
	Leu	Ser	**Stop**	Trp	G
C	Leu	Pro	His	Arg	U
	Leu	Pro	His	Arg	C
	Leu	Pro	Gln	Arg	A
	Leu	Pro	Gln	Arg	G
A	Ile	Thr	Asn	Ser	U
	Ile	Thr	Asn	Ser	C
	Ile	Thr	Lys	Arg	A
	Met	Thr	Lys	Arg	G
G	Val	Ala	Asp	Gly	U
	Val	Ala	Asp	Gly	C
	Val	Ala	Glu	Gly	A
	Val	Ala	Glu	Gly	G

REGULATION OF GENE EXPRESSION AND CELL SPECIALIZATION

What controls gene transcription, and how does an organism express only certain genes? Regulation of gene expression can occur at different times. The largest point is before transcription, or **pre-transcriptional regulation**. It can also occur post-transcriptionally or post-translationally.

The start of transcription requires DNA to be unwound and RNA polymerase to bind at the promoter. The process is usually a bit more complicated than that because there is a group of molecules called **transcription factors** that can either encourage or inhibit this from happening. This is often accomplished by making it easier or more difficult for RNA polymerase to bind or to move to the start site. The genes that are expressed cause observable cell differentiation which results in different phenotypes. The same transcription factors may influence different groups of genes.

Sometimes changes to the packaging of DNA will alter the ability of the transcription machinery to access a gene. These types of changes are called **epigenetic changes**, and they usually occur through a modification to a histone protein that is involved in winding up the DNA. A tighter wrap around a histone makes the DNA more difficult to access and a looser wrap makes it easier to access.

The following examples are famous models of regulation. You do not need to memorize them, but you should be familiar with the big picture of how things can come together to regulate transcription. Most of what we know about gene regulation comes from our studies of *E. coli*. In bacteria, a cluster of genes can be under the control of a single promoter; these functioning units of DNA are called **operons**. One of the best-understood operons is the *lac* operon, which controls expression of the enzymes that break down (catabolize) lactose.

The operon consists of four major parts: structural genes, promoter genes, the operator, and the regulatory gene:

- **Structural genes** code for enzymes needed in a chemical reaction. These genes will be transcribed at the same time to produce particular enzymes. In the *lac* operon, three enzymes (beta galactosidase, galactose permease, and thiogalactoside transacetylase) involved in digesting lactose are coded for.
- The **promoter gene** is the region where the RNA polymerase binds to begin transcription.
- The **operator** is a region that controls whether transcription will occur; this is where the repressor binds.
- The **regulatory gene** codes for a specific regulatory protein called the repressor. The repressor is capable of attaching to the operator and blocking transcription. If the repressor binds to the operator, transcription will not occur. On the other hand, if the repressor does not bind to the operator, RNA polymerase moves right along the operator and transcription occurs. In the *lac* operon, the **inducer**, lactose, binds to the repressor, causing it to fall off the operator, and "turns on" transcription.

The two diagrams below show the *lac* operon when lactose is absent (A) and when lactose is present (B).

Other operons, such as the *trp* operon, operate in a similar manner except that this mechanism is continually "turned on" and is "turned off" only in the presence of high levels of the amino acid tryptophan. Tryptophan is a product of the pathway that codes for the *trp* operon. When tryptophan combines with the *trp* repressor protein, it causes the repressor to bind to the operator, which turns the operon "off," thereby blocking transcription. In other words, a high level of tryptophan acts to repress the further synthesis of tryptophan.

The following two diagrams show the *trp* operon when tryptophan is absent (C) and when tryptophan is present (D).

Post-transcriptional regulation occurs when the cell creates an RNA, but then decides that it should not be translated into a protein. This is where RNAi comes into play. RNAi molecules can bind to an RNA via complementary base pairing. This creates a double-stranded RNA (remember RNA is usually single stranded). When a double-stranded RNA is formed, this signals to some special destruction machinery that the RNA should be destroyed. This would prevent it from going on to be translated.

Post-translational regulation can also occur if a cell has already made a protein, but doesn't yet need to use it. This is an especially common regulation for enzymes because when the cell needs them, it needs them ASAP. It is easier to make them ahead of time and then just turn them on or off as needed. This can involve binding with other proteins, phosphorylation, pH changes, cleavage, etc. Remember, even if a protein is made, it might need other things to be made (or not made) in order for the protein to be functional.

Gene Regulation in Embryonic Development

How does a tiny, single-celled egg develop into a complex, multicellular organism? By dividing, of course. The cell changes shape and organization many times by going through a succession of stages. This process is called **morphogenesis**.

When an egg is fertilized by a sperm, it forms a diploid cell called a **zygote**.

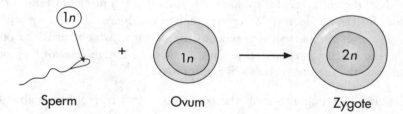

Fertilization triggers the zygote to go through a series of cell divisions. As these occur, the embryo becomes increasingly differentiated, or specialized. An undifferentiated cell is like a blank slate—it can become any type of cell. Once a cell begins to specialize, it is limited in its future options.

In order for a cell to differentiate, it must change. Certain genes will be expressed, and other genes might be turned off. Cells called organizers release signals that let each cell know how they should develop. As an over-simplified example, a cell destined to become a muscle might have an increase in things that make it flexible, while something destined to become bone might have a decrease in those things. Once the change has been made, the future muscle cell can't change its mind and become a bone cell. In other words, cells don't typically go backwards, or dedifferentiate.

The early genes that turn certain cells in the developing embryo into future-this or future-that are called **homeotic genes**. A subset of homeotic genes are called **Hox genes**. Timing is essential for the induction (activation) of these genes. Exactly the right bit of the embryo must be modified at exactly the right time. Otherwise, the embryo may develop with the brain in the wrong place, with too many limbs, or with limbs on only one side of the body. Severely damaged embryos will typically stop developing and die.

An interesting additional tool that the developing embryo uses is apoptosis, or programmed cell death. Certain bits of the developing embryo are used as a scaffold for development (like the cells that used to exist in the spaces between your fingers and toes), and when they are no longer necessary, those cells undergo apoptosis. Think of it like an eraser coming along and erasing the webbing tissue between fingers and toes.

MUTATIONS

A **mutation** is an error in the genetic code. Mutations can occur because DNA is damaged and cannot be repaired or because DNA damage is repaired incorrectly. Damage can be caused by chemicals or radiation. For example, radiation can cause breaks in the DNA strand. External chemicals can cause inappropriate chemical bonds to form between different parts of DNA, known as cross-linking. A mutation can also occur when a DNA polymerase or an RNA polymerase makes a mistake. DNA polymerases have proofreading abilities, but RNA polymerases do not. This is because RNA is only a temporary molecule and if a mistake is made, it is not usually as problematic. DNA is important now and in the future, since it is passed on from cell to cell in somatic cells and from parent to offspring. If the mistake occurs in a germline cell, that will become a gamete. Mistakes in DNA can last forever.

It is important to know that just having an error in the DNA is not a problem unless that gene is expressed AND the error causes a change in the gene product (RNA or protein). Think of it like an error in a recipe in a cookbook: an error won't harm anyone unless the recipe is actually made AND the error causes a big change in the recipe. As we will see, a mutation can cause different types of effects. Often, the result will be a change to the gene product (protein) and the phenotype of the individual caused by that protein. This can be beneficial, detrimental, or neutral to the individual depending on the specific mutation that occurs and the environmental context. Mutations are the primary source of genetic variation.

Base Substitution

Base substitution (point) mutations result when a single nucleotide base is substituted for another. There are three different types of point mutations:

There are two types of mutations: **base substitutions** and **gene rearrangements**.

- **Nonsense mutations** cause the original codon to become a stop codon, which results in early termination of protein synthesis.
- **Missense mutations** cause the original codon to be altered and produce a different amino acid.
- **Silent mutations** happen when a codon that codes for the same amino acid is created and therefore does not change the corresponding protein sequence.

Gene Rearrangements

Gene rearrangements involve DNA sequences that have deletions, duplications, inversions, and translocations.

1. **Insertions** and **deletions** result in the gain or loss, respectively, of DNA or a gene. They can involve either the addition (insertion) or removal (deletion) of as little as a single base or much larger sequences of DNA. Insertions and deletions may have devastating consequences during protein translation because the introduction or deletion of bases often results in a change in the sequence of codons used by the ribosome (called a **frameshift mutation**) to synthesize a polyprotein.

    Example:

Original DNA Sequence:	ATG	TAT	AAA	CAT	TGA
Original Polyprotein:	Met	Tyr	Lys	His	Stop
DNA Sequence after Insertion:	ATG	**C**TA	TAA	ACA	TTG A
Polyprotein Sequence after Insertion:	Met	Leu	Stop	-	-

    In this example, the insertion of an additional cytosine nucleotide resulted in a frameshift and a premature stop codon.

2. **Duplications** can result in an extra copy of genes and are usually caused by unequal crossing-over during meiosis or chromosome rearrangements. This may result in new traits because one copy of the gene can maintain the original function and the other copy may evolve a new function.

3. **Inversions** can result when changes occur in the orientation of chromosomal regions. This may cause harmful effects if the inversion involves a gene or an important regulatory sequence.

4. **Translocations** occur when two different chromosomes (or a single chromosome in two different places) break and rejoin in a way that causes the DNA sequence or gene to be lost, repeated, or interrupted.

5. **Transposons** are gene segments that can cut/paste themselves throughout the genome. The presence of a transposon can interrupt a gene and cause errors in gene expression.

Errors in Meiosis

Nondisjunction occurs when the chromosomes are not separated correctly during meiosis. This can cause entire chromosomes to be duplicated in a gamete, which can cause three copies of a chromosome (triploidy) in offspring. Incorrect numbers of chromosomes often lead to sterility and in plants can cause increased vigor due to polyploidy.

Bacteria

Bacteria, common pathogens, are prokaryotes that come in many shapes and sizes. They can infect many things, and sometimes they cause harm and sometimes they do not. You may have heard of "gut bacteria" before; this is a special colony of bacteria that lives inside each one of us. It helps us with some digestion and makes some things we need. We have a mutualistic relationship with our gut bacteria.

Bacteria divide by fission; however, this does not increase their genetic diversity. Instead, they can perform **conjugation** with other bacterial cells and swap some of their DNA. Bacteria can also gain genetic information through transformation (uptake of DNA) and transposition (movement of DNA within and between DNA molecules). All of these methods increase the genetic variation of bacteria. Genetic variety among bacteria is leading to increased antibiotic resistance.

Viruses

Viruses are nonliving agents capable of infecting cells. Why are viruses considered nonliving? They require a host cell's machinery in order to replicate. A virus consists of two main components: a protein shell (the capsid) and genetic material made of DNA or RNA, depending on the virus. Viruses are all very specific in which type of cells they infect, and the thing infected by a virus is called a **host.**

Viruses have one goal: replicate and spread. In order to do this, a virus needs to make more genome and make more capsid. They then assemble together into new viral particles.

The viral genome carries genes for building the capsid and anything else the virus needs that the host cannot provide. Sometimes, if two viruses infect the same cell, there will be mixing of the genomes, especially if the viruses have genomes split between several chromosome-like segments. A new virus particle might emerge that is a blend of the two viruses.

A commonly studied virus is a **bacteriophage** (a virus that infects bacteria). Bacteriophages undergo two different types of replication cycles, the **lytic cycle** and the **lysogenic cycle**. In the lytic cycle, the virus immediately starts using the host cell's machinery to replicate the genetic material and create more capsid proteins. These spontaneously assemble into mature viruses and cause the cell to lyse, or break open, releasing new viruses into the environment. In the lysogenic cycle, the virus incorporates itself into the host genome and remains dormant until it is triggered to switch into the lytic cycle. A virus can hide in the genome of a host cell for a very long time. During this time, the cell may divide and replicate the virus as well. By the time the lytic cycle is triggered, the virus may have been replicated many, many times as the cell hosting it divides.

The human microbiome is the collection of all the microbes in and on your body. We depend on a huge number of microbes to stay alive. Characterizing and understanding this population is a growing area of research.

Variations on Genetic Variation
Check out Chapter 10 for a more positive look at how an increase in genetic variation makes it more likely that a population will survive a catastrophic event.

When a virus excises from a host genome (becomes unintegrated), it sometimes accidentally takes some of the bacterial cell's DNA with it. Then, the host DNA accidentally gets packaged into new viral particles with the viral genome. The next cell that gets infected is not only getting infected with the viral genome, but also with that stolen chunk of bacterial DNA. If that chunk held a gene for something like antibiotic resistance, the next cell that gets infected will gain that trait. The transfer of DNA between bacterial cells using a lysogenic virus is called **transduction**.

Viruses that infect animals do not have to break their way out of the cell the same way that a bacteriophage does. Since animals' cells don't have a cell wall, viruses often just "bud" out of the membrane in a process similar to exocytosis. When a virus does this, it becomes enveloped by a chunk of host cell membrane that it takes with it. Viruses with a lipid envelope are called **enveloped viruses**.

Retroviruses like HIV are RNA viruses that use an enzyme called **reverse transcriptase** to convert their RNA genomes into DNA so that they can be inserted into a host genome. RNA viruses have extremely high rates of mutation because they lack proofreading mechanisms when they replicate their genomes. This high rate of mutation will create lots of variety, which makes these viruses difficult to treat. They evolve quickly, as drug-resistant mutations become naturally selected. New drugs must constantly be identified to treat the resistance.

Viruses can sometimes change quickly if two viruses infect the same cell. The gene segments of each virus can recombine and mix with each other and package into mixed viruses that are a chimera of the original two viruses. This can increase genetic variation among viruses.

BIOTECHNOLOGY

Recombinant DNA

Scientists have learned how to harness the Central Dogma in order to research and cure diseases. **Recombinant DNA** is generated by combining DNA from multiple sources to create a unique DNA molecule that is not found in nature. A common application of recombinant DNA technology is the introduction of a eukaryotic gene of interest (such as insulin) into a bacterium for production. In other words, bacteria can be hijacked and put to work as little protein factories. This branch of technology that produces new organisms or products by transferring genes between cells is called **genetic engineering**.

Polymerase Chain Reaction (PCR)

Polymerase chain reaction (PCR) is a laboratory technique that is used to create billions of identical copies of genes within hours. The process of creating many copies of genes is known as **amplification**. PCR utilizes a machine called a thermocycler to mimic the process of DNA replication, thereby creating many copies of genes. Amplified DNA fragments have many practical uses. For example, they can be utilized to perform phylogenetic analyses.

Transformation

Insulin, the protein hormone that lowers blood sugar levels, can now be made for medical purposes by bacteria. Yes, bacteria can be induced to use the universal DNA code to transcribe and translate a human gene! The process of giving bacteria foreign DNA is called **transformation**.

Not only has this laboratory technique been used to safely mass-produce important proteins used for medicine, such as insulin, but it also plays an important role in the study of gene expression. A technique similar to transformation is **transfection**, which is putting a plasmid into a eukaryotic cell, rather than a bacteria cell. This is a bit trickier than transformation. Because eukaryotic cells do not grow like bacteria do, transfection is not as useful for making large quantities of something, but sometimes it is important to use eukaryotic cells.

Gel Electrophoresis

An important part of DNA technology is the ability to observe differences in different preparations of DNA.

DNA fragments can be separated according to their molecular weight and charge with **gel electrophoresis**. Because DNA and RNA are negatively charged, they migrate through a gel toward the positive pole of the electrical field. The smaller the fragments, the faster they move through the gel. Restriction enzymes are also used to create a molecular fingerprint. The patterns created after cutting with restriction enzymes are unique for each person because each person's DNA is slightly different. Some people might have sequences that are cut many times, resulting in many tiny fragments, and others might have sequences that are cut infrequently, producing a few large fragments. When restriction fragments between individuals of the same species are compared, the fragments differ in length because of polymorphisms, which are slight differences in DNA sequences. These fragments are called **restriction fragment length polymorphisms**, or RFLPs. In **DNA fingerprinting**, RFLPs produced from DNA left at a crime scene are compared to RFLPs from the DNA of suspects.

DNA Sequencing

An important tool of modern scientists is the process of **DNA sequencing**. This allows scientists to determine the order of nucleotides in a DNA molecule. By knowing this, scientists could design their own DNA plasmid and use it to study a gene of interest.

CHAPTER 9 KEY TERMS

nucleotide
five-carbon sugar
phosphate
nitrogenous base
deoxyribose
adenine
guanine
cytosine
thymine
plasmids
phosphodiester bonds
double helix
Watson, Crick, and Franklin
base pairing
complementary
antiparallel
hydrogen bonds
gene
genome
chromosome
histone
nucleosome
euchromatin
heterochromatin
DNA replication
helicase
replication fork
origins of replication
topoisomerase
DNA polymerase
RNA primase
RNA primer
leading strand
lagging strand
Okazaki fragments
DNA ligase
semiconservative
telomeres
transcription
translation
Central Dogma of Biology
ribose
uracil
messenger RNA (mRNA)
ribosomal RNA (rRNA)
transfer RNA (tRNA)

RNA interference
   (RNAi)/silencing RNA (siRNA)
polycistronic transcript
monocistronic transcript
promoter
antisense/noncoding/minus/
   template strand
sense/coding strand
RNA polymerase
start site
exons
introns
splicing
spliceosome
poly(A) tail
5' GTP cap
codon
anticodon
wobble pairing
initiation
elongation
termination
A site, P site, E site
start codon
stop codons
pre-transcriptional regulation
transcription factors
epigenetic changes
operon
structural genes
promoter genes
operator
regulatory gene
inducer
post-transcriptional regulation
post-translational regulation
morphogenesis
zygote
fertilization
homeotic genes
Hox genes
mutation
base substitution
nonsense mutation
missense mutation
silent mutation

gene rearrangements
insertions
deletions
frameshift mutation
duplications
inversions
translocations
transposons
bacteria
conjugation
viruses
host
bacteriophage
lytic cycle
lysogenic cycle
transduction
enveloped virus
retrovirus
reverse transcriptase
recombinant DNA
genetic engineering
polymerase chain reaction (PCR)
amplification
transformation
transfection
gel electrophoresis
restriction fragment length poly-
   morphism (RFLP)
DNA fingerprinting
DNA sequencing

Chapter 9 Drill

Answers and explanations can be found in Chapter 15.

1 Mark for Review

DNA has to be replicated in all living individuals to copy DNA for biological inheritance. DNA polymerase is a crucial enzyme that catalyzes a reaction in the process. Which of the following best describes the reaction catalyzed by DNA polymerase?

(A) A phosphate group is added on to the three-prime end of a DNA strand.

(B) The double helix structure of the DNA strand is unwound.

(C) Nucleotides on the template strand are paired with complementary bases on the new strand.

(D) The sugar-phosphate bonds between the bases are broken.

2 Mark for Review

A researcher is preparing a sample of DNA for separation by electrophoresis. The researcher pipettes the DNA sample into a well in an agarose gel, which is then placed in an electrophoresis box and covered with a buffer. The box is connected to an electric current. The DNA sample fragments migrate through the gel toward the positive pole. Which of the following identifies the property of DNA that is responsible for the migration of the fragments?

(A) The negatively charged phosphate groups on DNA molecules

(B) The negatively charged polar bonds on the deoxyribose molecule

(C) The positively charged nitrogenous bases

(D) The negatively charged histone proteins that pack the DNA strands

3 Mark for Review

Kiwako Sakabe and Reiji Okazaki discovered Okazaki fragments in 1966 while researching DNA replication in *E. coli*. Which of the following correctly identifies the enzyme that connects Okazaki fragments to each other during DNA replication?

(A) DNA polymerase

(B) RNA polymerase

(C) Helicase

(D) DNA ligase

4 Mark for Review

People with cystic fibrosis tend to have thick secretions that cause coughs and lung infections and only affect those with two mutated copies of the cystic fibrosis gene. Those who are heterozygous for the allele do not experience the condition. The condition is typically caused when an amino acid at position 508 in the CFTR protein, a chloride ion channel, is removed through an error in the DNA replication process. Which of the following is the correct term for this type of mutation?

(A) Base insertion

(B) Base substitution

(C) Base deletion

(D) Silent mutation

AP Biology Premium Prep

5 Mark for Review

Some bacteria have a system that promotes mutation of DNA through recombination. Mutations allow different genes to be expressed and code for different proteins. The diagram below shows one rearrangement in the pattern of genes in one type of bacteria. Which of the following is the correct term for this type of rearrangement?

TUVWXYZ → TUVWVWXYZ

(A) Deletion

(B) Translocation

(C) Duplication

(D) Inversion

6 Mark for Review

Erythromycin is an antibiotic that blocks the growth of bacteria by binding to bacterial ribosomes and preventing the formation of peptide bonds. In some cells, the antibiotic does not affect all protein production. Which of the following explains this discrepancy?

(A) Protein synthesis is prevented from occurring in the cytoplasm.

(B) Peptide bonds are not necessary for all protein synthesis, so proteins are still able to form.

(C) Ribosomes do not play a role in protein synthesis, so proteins are still able to form.

(D) The proteins that are produced are translated from mRNA that was synthesized before erythromycin was applied.

7 Mark for Review

The firefly luciferase gene can be isolated and introduced to tobacco plants, resulting in light emission from the plants. Which of the following statements describes a necessary component of this occurrence?

(A) Tobacco plants and fireflies must be infected by the same pathogens.

(B) The processes of transcription and translation must be similar in both species.

(C) Chloroplasts must glow when exposed to firefly proteins.

(D) Fireflies and tobacco plants must have evolved from a relatively recent common ancestor.

8 Mark for Review

Leukemia in mice is caused by an oncoretrovirus, a retrovirus whose genetic information is inserted into its host's genome. Which of the following statements best explains how this process works?

(A) Reverse transcriptase converts the RNA from the retrovirus into DNA.

(B) DNA polymerase transcribes the retrovirus's RNA into DNA.

(C) The retrovirus's RNA is similar enough to DNA to be incorporated into the mouse genome without the aid of any enzymes.

(D) DNA ligase attaches DNA nucleotides to the retrovirus's RNA, which allows it to link to the mouse's genome.

230 | For more free content, visit PrincetonReview.com

AP Biology Premium Prep

9 Mark for Review

Why do individual cells of an organism produce different proteins during embryonic development?

(A) The autosomal cells vary in their number of chromosomes.

(B) Not all cells synthesize working proteins.

(C) Some genes are deleted from the genome as cells differentiate.

(D) Some cells have different genes activated than other cells.

10 Mark for Review

A cell's unique role in a living organism is primarily influenced by which of the following?

(A) Nucleotide sequence in DNA

(B) Reagents that restrict the formation of microtubules

(C) Nuclear gene expression

(D) Molecules in cytoplasm

11 Mark for Review

The polymerase chain reaction is often used in molecular biology studies. Which of the following describes a benefit of this procedure?

(A) It can produce a large quantity of a desired DNA segment relatively rapidly.

(B) It produces flawless copies of DNA segments using cheap and accessible materials.

(C) It purifies specific DNA segments from other DNA segments.

(D) It duplicates the entire genome of the individual being analyzed.

9. Gene Expression and Regulation | **231**

Chapter 9 Summary

- o DNA is the genetic material of the cell.

- o DNA structure is composed of two antiparallel complementary strands, including:
  - Deoxyribose sugar and phosphate backbone connected by phosphodiester bonds, with nitrogenous bases hydrogen bonded together pointing toward the middle. The two strands are oriented antiparallel with 3' and 5' ends and twist into a double helix.
  - Adenine always pairs with thymine, and guanine always pairs with cytosine.

- o All the DNA in a cell is the genome. It is divided up into chromosomes, which are divided up into genes that each encode a particular genetic recipe.

- o DNA replication provides two copies of the DNA for cell division, which involves the following steps:
  - Helicase begins replication by unwinding DNA and separating the strands at the origin of replication.
  - Topoisomerases reduce supercoiling ahead of the replication fork.
  - RNA primase places an RNA primer down on the template strands.
  - DNA polymerase reads DNA base pairs of a strand as a template and lays down complementary nucleotides to generate a new complementary partner strand.
  - Replication proceeds continuously on the leading strand.
  - The lagging strand forms Okazaki fragments, which must be joined later by DNA ligase.

- o Transcription involves building RNA from DNA.
  - RNA polymerase reads one of the DNA strands (the antisense/noncoding/template strand) and adds RNA nucleotides to generate a single strand of mRNA that will be identical to the other strand of DNA (the coding/sense strand) except it will have uracil instead of thymine.
  - mRNA, rRNA, tRNA, and RNAi are all types of RNA that can be created.
  - mRNA gets processed in eukaryotic cells: introns are removed (spliced out), and a 5' GTP cap and a 3' poly(A) tail are added.

- Translation occurs in the cytoplasm as mRNA carries the message to the ribosome.
    - mRNA passes through three sites on the ribosome, the A-site, the P-site, and the E-site.
    - The mRNA is read in triplets of three nucleotides, called codons. Each tRNA has a region called the anticodon that is complementary to a codon. Sometimes the pairing is not the normal base-pairing. This is called wobble pairing.
    - When a tRNA binds, it brings the corresponding amino acid to add to the growing polypeptide.

- Gene expression is regulated primarily by transcription factors influencing transcription (pre-transcriptional regulation), RNAi after transcription (post-transcriptional regulation), and by various reactions and regulators after translation (post-translational regulation). This regulation is dynamic and can either increase or decrease gene expression, RNA levels, and protein levels according to the needs of the cell.

- Mutations can result from changes in the DNA message or the mRNA message.

- Mutations can be small (single nucleotide swaps, additions, or deletions) or large (big chunks or entire chromosomes are swapped, duplicated, or deleted).

- Some examples of biotechnology are:
    - recombinant DNA
    - polymerase chain reaction (PCR)
    - transformation of bacteria
    - gel electrophoresis

- Bacteria and viruses are common pathogens. Bacteriophages are viruses that infect bacteria. Viruses require a host to replicate and sometimes lyse the host cell during infection.

Chapter 10
Natural Selection

All of the organisms we see today arose from earlier organisms. This process, known as **evolution**, can be described as a change in a population over time. Interestingly, however, the driving force of evolution, called **natural selection**, operates on the level of the individual. In other words, evolution is defined in terms of populations but occurs in terms of individuals.

> Natural selection acts on individuals, but evolution acts on populations.

NATURAL SELECTION

What is the basis of our knowledge of evolution? Much of what we now know about evolution is based on the work of **Charles Darwin**. Darwin was a 19th-century British naturalist who sailed the world in a ship named the HMS *Beagle*. Darwin developed his theory of evolution based on natural selection after studying animals in the Galápagos Islands and other places.

He observed that there were similar animals on the various islands, but the beaks varied in length on finches, and necks varied in length on tortoises. There must be a reason why animals in different areas had different traits. Darwin concluded that it was impossible for the finches and tortoises of the Galápagos simply to "grow" longer beaks or necks as needed. Rather, those traits were inherited and passed on from generation to generation. So, he decided there must be a way for populations to evolve and change their traits (i.e., a population of finches developing longer beaks).

He decided that there must have been a variety of beak lengths originally, but only the longest one was particularly helpful. Since those finches could eat better, survive better, and reproduce better, they were more likely to contribute offspring to the next generation. Over many years, long-beak finches became more and more plentiful with each generation until long beaks were the "norm." In another example, on the first island Darwin studied, there must once have been short-necked tortoises. Unable to reach higher vegetation, these tortoises eventually died off, leaving only those tortoises with longer necks. Consequently, evolution has come to be thought of as "the survival of the fittest": only those organisms most fit to survive will survive and reproduce.

Darwin elaborated his theory in a book entitled *On the Origin of Species*. In a nutshell, here's what Darwin observed:

- Each species produces more offspring than can survive.
- These offspring compete with one another for the limited resources available to them.
- Organisms in every population vary.
- Evolutionary fitness is measured by reproductive success.
- The fittest offspring, or those with the most favorable traits, are the most likely to survive and therefore produce a second generation. These are the traits most likely to pass to subsequent generations.
- Fitness varies based on biotic and abiotic factors; different genetic variations can be selected for in different generations.

Lamarck and the Long Necks

Darwin was not the first to propose a theory explaining the variety of life on Earth. One of the most widely accepted theories of evolution in Darwin's day was that proposed by **Jean-Baptiste de Lamarck**.

In the 18th century, Lamarck had proposed that *acquired* traits were inherited and passed on to offspring. For example, in the case of Lamarck's giraffes, Lamarck's theory said that the giraffes had long necks because they were constantly reaching for higher leaves while feeding. This theory is referred to as the "law of use and disuse," or, as we might say now, "use it or lose it." According to Lamarck, giraffes have long necks because they constantly *use* them.

We know now that Lamarck's theory was wrong: acquired changes—that is, changes at a "macro" level in an organism's regular (somatic) body cells—will not appear in gamete cells. For example, if you were to lose one of your fingers, your children would not inherit this trait. The gametes that your body makes include copies of your regular old genome, not a new version of a genome that is determined by how careful you are with a table saw.

> **Don't be Lamarck!**
> Remember, Lamarck was wrong. Do not take a Lamarckian position on the AP Exam.

Evidence for Evolution

In essence, nature "selects" which living things survive and reproduce. Evidence supports the theory that all living organisms share a common ancestor. Today, we find support for the theory of evolution in several areas:

- **Paleontology**, or the study of fossils: paleontology has revealed to us both the great variety of organisms (most of which, including trilobites, dinosaurs, and the woolly mammoth, have died off) and the major lines of evolution. Fossils can be dated by:

  i. The age of the rocks where a fossil is found
  ii. The rate of decay of isotopes including carbon-14
  iii. Geographical data

- **Biogeography**, or the study of the distribution of **flora** (plants) and **fauna** (animals) in the environment: scientists have found related species in widely separated regions of the world. For example, Darwin observed that animals in the Galápagos have traits similar to those of animals on the mainland of South America. One possible explanation for these similarities is a common ancestor. As we'll see below, there are other explanations for similar traits. However, when organisms share multiple traits, it's pretty safe to say that they likely also shared a common ancestor.

- **Embryology**, or the study of development of an organism: if you look at the early stages in vertebrate development, all embryos look alike! For example, all vertebrates—including fish, amphibians, reptiles, birds, and even mammals such as humans—show fishlike features called gill slits.

- **Morphological homologies**, or the study of the anatomy of various animals: scientists have discovered that some animals have similar structures that serve different functions. For example, a human's arm, a dog's leg, a bird's wing, and a whale's fin are all the same appendages, though they have evolved to serve different purposes. These structures, called **homologous structures**, also point to a common ancestor.

Human's arm   Dog's leg   Bird's wing   Whale's fin

In contrast, sometimes animals have features with the same function but that are structurally different. A bat's wing and an insect's wing, for example, are both used to fly. They therefore have the same function, but they have evolved totally independently of one another. These are called **analogous structures**. Another classic example of an analogous structure is the eye. Though scallops, insects, and humans all have eyes, these three different types of eyes are thought to have evolved entirely independently of one another. They are therefore analogous structures.

- **Molecular Biology:** Perhaps the most compelling proof of all is the similarity at the molecular level. Today, scientists can examine nucleotide and amino acid sequences of different organisms. From these analyses, we've discovered that organisms that are closely related have a greater proportion of sequences in common than distantly related species. For example, humans don't look much like chimpanzees. However, by some estimates, as much as 99 percent of our genetic code is identical to that of a chimp. All eukaryotes have membrane-bound organelles, linear chromosomes, and genes that contain introns.

- **Continuing Evolution:** There is evidence that evolution is constantly occurring. We can see consistent small changes in DNA and changes in the fossil record. We can see evolution of resistance to antibiotics, pesticides, herbicides, or chemotherapy drugs since these are particularly strong selective pressures. We can also see how fast replicating pathogens evolve and cause emergent diseases never before seen.

Across every domain, there are certain universal molecules and structures. Many metabolic pathways are also conserved and vary only a little between species.

COMMON ANCESTRY

By using the evidence mentioned above, scientists can get a good handle on how evolution of certain species occurred. Structural evidence indicates common ancestry of all living eukaryotes based on membrane-bound organelles, linear chromosomes, and genes that contain introns. It all comes down to who has common ancestors. Life started somewhere. Some original life-form is the **common ancestor** to all life, but where things went from there can become quite convoluted.

Scientists use charts called **phylogenetic trees,** or **cladograms**, to study relationships between organisms. Phylogenetic trees are built using data from the fossil record or molecular record, but the molecular record provides more specific details. Trees show the amount of change over time, but cladograms just show the change. In other words, cladograms are often drawn with even spacing between species, but phylogenetic trees are often drawn with different distances between species and as a result they look more like a tree with uneven branches. Both types are hypotheses and thus they constantly change as more data is uncovered to establish the relationships between species.

They always begin with the common ancestor and then branch out. Anytime there is a fork in the road, it is called a common ancestor node. These common ancestors likely do not exist anymore, but they are the point at which evolution went in two directions. One direction eventually led to one species, and the other eventually led to another species.

For example, we have a common ancestor with chimps. This doesn't mean that our great-great...great-grandfather was a chimp. It means that chimps and we have the same great-great...great-grandfather, who was neither a human nor a chimp. Instead, he was some other species completely, and a speciation event must have occurred where his species was split and evolved in different directions. One lineage became chimps, and the other linage became humans.

This is an example of a phylogenetic tree.

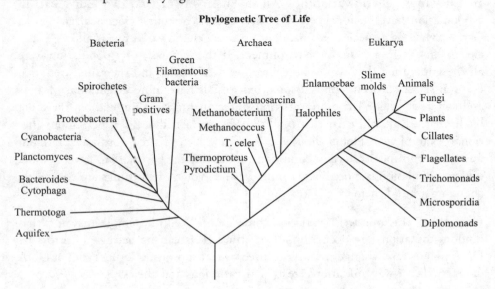

Phylogenetic Tree of Life

In this tree, the three main domains of life (Bacteria, Archaea, and Eukarya) are seen. You can see that plants, animals, and fungi are much more closely related to each other than they are to bacteria. Archaea and eukarya have a common ancestor with bacteria, and then they also have another common ancestor that bacteria don't have.

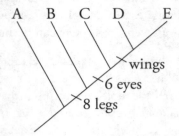

In the phylogenetic tree above, species A is an **out-group** because it is the least related to all of the other species. Traits that are either gained or lost are used to construct phylogenetic trees. Shared traits indicate common ancestry. Phylogenetic trees are hypotheses and are constantly being revised as new evidence emerges. Species B, C, D, and E all have the shared character of 8 legs. Wings are a shared character for species D and E, but they could also be considered a derived character since they separate species D and E from the rest of the species that do not have wings. Populations of organisms are still evolving today. Genomes and the fossil record continue to change, prokaryotes develop resistance to antibiotics, pesticides, herbicides, or chemotherapy drugs, and pathogens evolve and cause emergent diseases.

GENETIC VARIABILITY

As you know, no two individuals are identical. The differences in each person are known as **genetic variability**. All this means is that no two individuals in a population have identical sets of alleles (except, of course, identical twins). In fact, survival of a species is dependent on this genetic variation; it allows a species to survive in a changing environment. If the members of a population were all the same, then they would have the same weaknesses and the same strengths. Natural selection occurs only if some individuals have more evolutionary fitness and can be selected. The more variations there are among a population, the more likely that a trait will exist that might be the perfect lifesaver. One famous (fictional) example is Rudolph the Red-Nosed Reindeer. His red nose was a random mutation, but under the right conditions, it became the best nose to have. A more diverse population is more likely to survive and evolve when things are constantly changing around them.

How did all this wonderful variation come about? Well, the simple answer is that random mutations are occurring all the time. This can be because of errors by DNA polymerase, changes to DNA caused by transposons, or other types of DNA damage. Either way, mutations create new variations and alleles.

However, just the occurrence of mutations does not account for the great amount of genetic variation in a population. The mixing of genes through sexual reproduction also contributes to genetic variation. During meiosis, crossing-over mixes alleles among homologous chromosomes, and independent assortment when chromosomes are packaged further adds to the genetic uniqueness of each gamete.

It might be hard to think of it in this way, but genetic variation is the very foundation of evolution, as we'll soon see. Now that we've reintroduced genes, we can refine our definition of evolution. More specifically:

> Evolution is the change in the gene pool of a population over time.

The Peppered Moths

Let's look at an example. During the 1850s in England, there was a large population of **peppered moths**. In most areas, exactly half of them were dark and carried alleles for dark coloring. The other half were light and carried alleles for light coloring. This 1:1 ratio of phenotypes was observed until air pollution, due primarily to the burning of coal, changed the environment. What happened?

Imagine two different cities: City 1 (in the south) and City 2 (in the north). Prior to the Industrial Revolution, both of these cities had unpolluted environments. In both of these environments, dark moths and light moths lived comfortably side by side. For simplicity's sake, let's say their proportions were a perfect fifty-fifty, half dark and half light. However, at the height of the Industrial Revolution, City 2, our northern city, was heavily polluted, whereas City 1, our southern city, was unchanged. In the north, where all the trees and buildings were covered with soot, the light moths didn't stand a chance. They were impossible for a predator to miss! As a result, the predators gobbled up light-colored moths just as fast as they could reproduce, sometimes even before they reached an age in which they *could* reproduce. However, the dark moths with their dark alleles were just fine. With all the soot around, the predators couldn't even see them; the dark moths continued doing their thing—above all, *reproducing*. And when they reproduced, they had more and more offspring carrying the dark allele.

After a few generations, the peppered moth gene pool in City 2 changed. Although our original moth gene pool was 50 percent light and 50 percent dark, excessive predation changed the population's genetic makeup. By about 1950, the gene pool reached 90 percent dark alleles and only 10 percent light alleles. This occurred because the light moths didn't stand a chance in an environment where they were so easy to spot. The dark moths, on the other hand, multiplied just as fast as they could.

In the southern city, you'll remember, there was very little pollution. What happened there? Things remained pretty much the same. The gene pool was unchanged, and the population continued to have roughly equal proportions of light moths and dark moths.

CAUSES OF EVOLUTION

Natural selection, the evolutionary mechanism that "selects" which members of a population are best suited to survive and which are not, works both "internally" and "externally": *internally* through random mutations and *externally* through environmental pressures. Biotic and abiotic factors can fluctuate, and this can affect the direction of evolution. With each generation, different traits might be selected for depending on environmental fluctuations.

> Natural selection requires genetic variation and an **environmental pressure** that gives some individuals an advantage.

To see how this process unfolds in nature, let's return to the moth case. Why did the dark moths in the north survive? Because they were dark-colored. But how did they become dark-colored? The answer is, through **random mutation**. One day, long before the coal burning, a moth was born with dark-colored wings. As long as a mutation does not kill an organism before it reproduces (most mutations, in fact, do), it may be passed on to the next generation. Over time, this one moth had offspring. These, too, were dark. The dark- and the light-colored moths lived happily side by side until something from the outside—in our example, the environment—changed all that.

The initial variation came about by chance. This variation gave the dark moths an edge. However, that advantage did not become apparent until something made it apparent. In our case, that something was the intensive pollution from coal burning. The abundance of soot made it easier for predators to spot the light-colored moths, thus effectively removing them from the population. Therefore, dark color is an **adaptation**, a variation favored by natural selection.

Survival of the fittest is the name of the game, and any trait that causes an individual to reproduce better gives that individual **evolutionary fitness**.

These traits are often things that simply help something to survive. After all, survival is essential for reproduction to occur. Strength, speed, height, camouflage, and many other things can be helpful. Sometimes odd things can be helpful as well. For example, a peacock's tail has been selected for because females choose to mate with males that have a large and beautiful tail. The tail doesn't help them to survive because it actually makes it easier for predators to catch them, but it is essential for the females to find them attractive. This is an example of **sexual selection**.

Let's go back to our moth example. We ended up with different gene pools in the two populations (north vs. south). Eventually, over long stretches of time, these two different populations might change so much that they could no longer reproduce together. At that point, we would have two different species, and we could say, definitively, that the moths had evolved. Evolution occurred as a consequence of random mutation and the pressure put on the population by an environmental change. Catastrophic events can speed up natural selection and adaptation. When the rules change, sometimes odd variations are selected for. Earth has undergone several mass extinction events. The tree of life would look much different if these events had not occurred.

Fittest of the Fit

Don't confuse evolutionary fitness with physical fitness. Often, things like strength and speed will increase evolutionary fitness, but anything that increases survival and reproducing will add to evolutionary fitness, even if sometimes those individuals are slow or weak. Keep in mind that fitness depends on the environment. A trait may be beneficial in one circumstance and detrimental in another. Selection takes place one moment at a time in one environment at a time.

Phenotypes Not Genotypes

Remember, it is a phenotype that can be selected for/against. It is not the genotype. Genotypes are hidden away inside the DNA; only the phenotype will be exposed to the environmental pressures.

several mass extinction events. The tree of life would look much different if these events had not occurred.

Genetic drift is something that causes a change in the genetics of a population, but it is not natural selection. Instead, it is caused by random events that drastically reduce the number of individuals in a population. Often called the **bottleneck effect** or the **founder effect**, this occurs when only a few individuals are left to mate and regrow a population. If those individuals were the only survivors due to random luck, that means that the alleles that are present to regrow the population are just random alleles. They are not necessarily the fittest alleles. For example, if a hurricane swept through and only four birds were left on an island, the traits that they have will be the traits of the population going forward, even if those traits were not previously being selected for.

Gene flow can occur between different populations of the same species if individuals migrate to/from the populations. In other words, new alleles are entering/exiting and the genetic diversity will change. Small populations are more susceptible than large populations to this type of change.

Types of Selection

The situation with our moths is an example of **directional selection**. One of the phenotypes was favored at one of the extremes of the normal distribution.

In other words, directional selection "weeds out" one of the phenotypes. In our case, dark moths were favored, and light moths were practically eliminated. Here's one more thing to remember: directional selection can happen only if the appropriate allele—the one that is favored under the new circumstances—is already present in the population.

Two other types of selection are **stabilizing selection** and **disruptive selection**. Stabilizing selection means that organisms in a population with extreme traits are eliminated. This type of selection favors organisms with average or medium traits. It "weeds out" the phenotypes that are less adaptive to the environment. A good example is birth weight in human babies. An abnormally small baby has a higher chance of having birth defects; conversely, an abnormally large baby will have a challenge in terms of a safe birth delivery.

Disruptive selection, on the other hand, does the reverse. It favors both the extremes and selects *against* common traits. For example, females are "selected" to be small, and males are "selected" to be large in elephant seals. You'll rarely find a female or male of intermediate size.

Types of Selection

Another type of selection to be aware of is **artificial selection**. This is a type of selection where humans directly affect variation in other species. For example, colors of plants and sizes of dogs are greatly influenced by humans who are orchestrating breeding to create traits that we desire. This makes these traits selected for even if they are not increasing the reproductive fitness of a species.

SPECIES

What Is a Species?
To be considered the same species, two individuals must be able to mate and produce viable offspring that would then be capable of mating and producing offspring.

A dog and a bumblebee obviously cannot produce offspring. Therefore, they are different **species**. However, a chihuahua and a Great Dane can reproduce (at least in theory). We would not say that they are different species; they are merely different breeds. In order for these to become different species of dogs, they would have to become **reproductively isolated** from each other. This would then allow the two groups to undergo natural selection and evolve differently. With different variation and different environmental pressures, they could each change in different ways and no longer be able to mate. This is called **divergent evolution**. Divergent evolution that occurs quickly after a period of stasis (little evolution) is called **punctuated equilibrium**. This is often due to a major event. A change that comes about after many smaller changes is called **gradualism**, and it takes hundreds or millions of years. You should also be aware of **adaptive radiation**, which occurs when a species rapidly diversifies due to an abundance of available ecological niches suddenly opening up.

Divergent evolution results in closely related species with different behaviors and traits. As with our example, these species often originate from a common ancestor. More often than not, the "engine" of evolution is cataclysmic environmental change, such as pollution in the case of the moths. Geographic barriers, new stresses, disease, and dwindling resources are all factors in the process of evolution. Pre- and post-zygotic barriers also prevent organisms of two different species from mating to produce viable offspring. **Pre-zygotic barriers** prevent fertilization. Examples of this kind of barrier include temporal isolation, which occurs when two species reproduce at different times of the year. A **post-zygotic barrier** is related to the inability of a hybrid organism to produce offspring. For example, a horse and a donkey can mate to produce a mule, but mules are sterile and therefore cannot produce a second generation.

Convergent evolution is the process by which two unrelated and dissimilar species come to have similar (analogous) traits, often because they have been exposed to similar selective pressures. Examples of convergent evolution include aardvarks, anteaters, and pangolins. They all have strong, sharp claws and long snouts with sticky tongues to catch insects, yet they evolved from three completely different mammals.

There are two types of **speciation**: **allopatric speciation** and **sympatric speciation**. Allopatric speciation simply means that a population becomes separated from the rest of the species by a geographic barrier so that the two populations can't interbreed. An example would be a mountain that separates two populations of ants. In time, the two populations might evolve into different species. If, however, new species form without any geographic barrier, it is called sympatric speciation. This type of speciation is common in plants. Two species of plants may evolve in the same area without any geographic barrier. Often plants undergo speciation due to **polyploidy**, which occurs when they get doubles of their chromosomes.

POPULATION GENETICS

Mendel's laws can extend to the population level. Suppose you caught a bunch of fruit flies—about 1,000. Let's say that 910 of them were red-eyed and 90 were green-eyed. If you allowed the fruit flies to mate and counted the next generation, we'd see that the ratio of red-eyed to green-eyed fruit flies would remain approximately the same: 91 percent red-eyed and 9 percent green-eyed. That is, the allele frequency would remain constant. At first glance you may ask, how could that happen?

Hardy-Weinberg Equilibrium

Hardy-Weinberg is a model for describing and predicting allele frequencies in a nonevolving population. The **Hardy-Weinberg law** states that even with all the shuffling of genes that goes on, the relative frequencies of genotypes in a population are constant over time. Alleles don't get lost in the shuffle. The dominant gene doesn't become more prevalent, and the recessive gene doesn't disappear.

This law says that a population will be in genetic equilibrium only if it meets these five conditions: (1) a large population, (2) no net mutations, (3) no immigration or emigration, (4) random mating, and (5) no natural selection.

Hardy-Weinberg in Real World Conditions

When the five conditions of Hardy-Weinberg equilibrium are met, the gene pool in a population is relatively stable. However, in real-world populations, it is unlikely that a population will meet all of the five conditions.

It is important to understand the consequences on the population if any of the five conditions are not met:

1. If the population is small, the population will be more susceptible to random environmental impact than if the population is large. A random process that causes allele frequencies to change is known as genetic drift. In other words, small populations are more susceptible to genetic drift.

2. If mutations are present in a population, new alleles will be introduced into the population and hence genetic equilibrium will be disturbed. For example, an error in DNA replication can result in a new allele being introduced.

3. If immigration or emigration is present, species entering or leaving a population will bring or remove alleles with them.

4. If there is non-random mating, individuals will be selective in how they pick their mating partners based on a trait or traits. For example, if female peahens prefer peacocks with brighter tails, the alleles that contribute to the brighter tails phenotype will be preferentially propagated, and allele frequencies will accordingly shift.

5. If there is natural selection, organisms better adapted to their environment will be more likely to survive and reproduce, and thus their alleles will be preferentially propagated to the next generation. Changes in allele frequencies provide evidence for evolution occurring in a population.

Hardy-Weinberg Equations

Let's say that the allele for red eyes, R, is dominant over the allele for green eyes, r. Red-eyed fruit flies include homozygous dominants, RR, and heterozygous, Rr. The green-eyed fruit flies are recessive, rr. The frequency of each allele is described in the equation below. The allele must be either R or r. Let "p" represent the frequency of the R allele and "q" represent the frequency of the r allele in the population.

$$p + q = 1$$

This sum of the frequencies must add up to one. If you know the value of one of the alleles, then you'll also know the value of the other allele. This makes sense because Hardy-Weinberg works ONLY for populations with 2 alleles that have normal dominant-recessive behavior. So, if you know there are only 2 alleles possible and you know that 70% are the dominant allele, then the remaining 30% must be the recessive allele.

We can also determine the frequency of the *genotypes* in a population using the following equation.

$$p^2 + 2pq + q^2 = 1$$

In this equation, p^2 represents the homozygous dominants, $2pq$ represents the heterozygotes, and q^2 represents the homozygous recessives.

Both of these equations are listed on the AP Biology Equations and Formulas sheet. But how do you use them? Use proportions in the population to figure out both the allele and genotype frequencies. Let's calculate the frequency of the genotype for green-eyed fruit flies. If 9% of the fruit flies are green-eyed, then the *genotype* frequency, q^2, is 0.09. You can now use this value to figure out the frequency of the recessive allele in the population. The allele frequency for green eyes is equal to the square root of 0.09—that's 0.3. If the recessive allele is 0.3, the dominant allele must be 0.7. That's because 0.3 + 0.7 equals 1.

Using the second equation, you can calculate the genotypes of the homozygous dominants and the heterozygotes. The frequency for the homozygous dominants, p^2, is 0.7 × 0.7, which equals 0.49. The frequency for the heterozygotes, $2pq$, is 2 × 0.3 × 0.7, which equals 0.42. If you include the frequency of the recessive genotype—0.09—the numbers once again add up to 1.

Extinction

There have been many extinctions throughout Earth's history. The rate of extinction can vary and is usually accelerated during ecological stress. Humans have been a large ecological stressor and have likely contributed to extinction events. A population with high genetic diversity is more protected from extinction events because there is a higher chance that an individual will have the characteristics required to survive the ecological stress leading to the extinction event. When a species becomes extinct, the niche it occupied can become available to other species.

ORIGINS OF LIFE ON EARTH

The Earth formed approximately 4.6 billion years ago. Prior to 3.9 billion years ago, the Earth was likely an extremely hostile place. The earliest evidence of life is approximately 3.5 billion years ago, but how did life arise? This is still a hotly debated topic among scientists. Most scientists believe that the earliest precursors of life arose from nonliving matter (basically, gases) in the primitive oceans of the earth. But this theory didn't take shape until the 1920s. Two scientists, **Alexander Oparin** and **J. B. S. Haldane**, proposed that the primitive atmosphere contained mostly inorganic molecules and was rich in the following gases: methane (CH_4), ammonia (NH_3), hydrogen (H_2), and water (H_2O). Interestingly enough, there was almost no free oxygen (O_2) in this early atmosphere. They believed that these gases collided, producing chemical reactions that eventually led to the organic molecules we know today.

Don't Forget the Heterozygotes!
Remember that the number of individuals with the dominant phenotype includes those with the homozygous dominant genotype and those with the heterozygous genotype.

Alternative Hypothesis
An alternative hypothesis is that organic molecules were transported to Earth on a meteorite or due to some other celestial event.

This theory didn't receive any substantial support until 1953. In that year, **Stanley Miller** and **Harold Urey** simulated the conditions of primitive Earth in a laboratory. They put the gases theorized to be abundant in the early atmosphere into a flask, struck them with electrical charges in order to mimic lightning, and organic compounds similar to amino acids appeared!

But how do we make the leap from simple organic molecules to more complex compounds and life as we know it? Since no one was around to witness the process, no one knows for sure how (or when) it occurred. It is likely that the original life-forms were simply molecules of RNA. This is called the **RNA-world hypothesis**. RNA can take many shapes because it is not restricted to being a double helix. It is possible that RNA molecules capable of replicating and thus passing along themselves (their genome) were the first life-forms. Complex organic compounds (such as proteins) must have formed via dehydration synthesis. Simple cells then used organic molecules as their source of food. Over time, simple cells evolved into complex cells.

CHAPTER 10 KEY TERMS

evolution
natural selection
Charles Darwin
Jean-Baptiste de Lamarck
paleontology
biogeography
flora
fauna
embryology
morphological homologies
homologous structures
analogous structures
molecular biology
continuing evolution
common ancestor
phylogenetic tree (cladogram)
out-group
genetic variability
peppered moths
environmental pressure
random mutation
adaptation
evolutionary fitness
sexual selection
genetic drift

bottleneck (founder effect)
gene flow
directional selection
stabilizing selection
disruptive selection
artificial selection
species
reproductively isolated
divergent evolution
punctuated equilibrium
gradualism
adaptive radiation
pre-zygotic barriers
post-zygotic barriers
convergent evolution
speciation
allopatric speciation
sympatric speciation
polyploidy
Hardy-Weinberg law
Alexander Oparin and J. B. S. Haldane
Stanley Miller and Harold Urey
RNA-world hypothesis

AP Biology Premium Prep

Chapter 10 Drill

Answers and explanations can be found in Chapter 15.

1 ☐ Mark for Review

The Death's-head hawkmoth has markings on its back that resemble a bee, which deters predators. It also emits an odor that resembles that of a honey bee and allows it to enter a beehive and consume honey without being attacked by honey bees. Which of the following strategies is the Hawkmoth relying on for survival?

(A) Mimicry

(B) Coevolution

(C) Divergent evolution

(D) Convergent evolution

2 ☐ Mark for Review

Isolation prevents breeding between species based on differences in behavior, genetics, or morphology. One type of isolation can prevent energy and resources from being wasted when individuals produce gametes and zygotes that don't result in a viable offspring. Which of the following types of isolation occurs after a zygote has been produced?

(A) Temporal

(B) Mechanical

(C) Developmental

(D) Geographic

3 ☐ Mark for Review

Tasmanian wolves and native cats are marsupials found in Australia. Wolves and ocelots are placental animals found on other continents besides Australia that exhibit similar characteristics to Tasmanian wolves and native cats. Which of the following likely explains the similarities between the species?

(A) The species are exposed to comparable gene pools.

(B) The species follow comparable mating seasons.

(C) The species experience comparable predator-prey interactions.

(D) The species are exposed to comparable selection pressures.

4 ☐ Mark for Review

A polyploid individual within a plant population of diploid individuals is a possible new species. If the polyploid individual is capable of reproducing to form a new population, what would scientists consider this to be an example of?

(A) X-linked inheritance

(B) Sympatric speciation

(C) Genetic drift

(D) Allopatric speciation

250 | For more free content, visit PrincetonReview.com

AP Biology Premium Prep

5 ☐ Mark for Review

A population of frogs contains individuals of varying sizes. A researcher notices that large male frogs in this population successfully mate with females significantly more often than small male frogs do. Which of the following best explains this observation?

Ⓐ Small female frogs are more likely to mate with small male frogs, and large female frogs are more likely to mate with large male frogs.

Ⓑ The mating chirps produced by small male frogs are not heard by female frogs.

Ⓒ Small male frogs do not jump as high as large male frogs.

Ⓓ Large male frogs are more successful than small male frogs when competing for mates due to securing a larger breeding territory.

6 ☐ Mark for Review

Evolution does not occur when a population is said to be in equilibrium by the Hardy-Weinberg definition. Which of the following suggests that evolution will occur in a population?

Ⓐ The population is very large.

Ⓑ Natural selection pressures do not exist.

Ⓒ Individuals usually mate randomly.

Ⓓ Mutations occur in the population.

7 ☐ Mark for Review

A group of people living in an isolated area exhibits a higher incidence of a disease than the rest of the world. All of the people who exhibit this disease can track their descent back to one couple who were the original settlers of the area. What is the remarkably high incidence of this disease an example of?

Ⓐ Genetic equilibrium and sexual selection

Ⓑ The founder effect and genetic drift

Ⓒ Natural selection and mutation

Ⓓ The bottleneck effect and sexual selection

8 ☐ Mark for Review

Antibiotics have been used for decades to treat humans and combat infections caused by bacteria. However, some outbreaks of infections do not respond to antibiotics that were previously successful, causing a significant number of deaths. Which of the following could explain this discrepancy?

Ⓐ Bacteria that are resistant to antibiotics survive and multiply, which causes additional cases.

Ⓑ Bacteria from other organisms pass on new illnesses to humans that are not affected by the antibiotics.

Ⓒ The bacteria mutated to be able to use the antibiotics as an energy source.

Ⓓ The bacteria evolved strategies to avoid the antibiotics when they are prescribed to humans.

9 ☐ Mark for Review

In a certain chicken coop, 5% of the population has gray feathers and 95% of the population has white feathers. The population is in Hardy-Weinberg equilibrium. If a farmer sells some of the chickens with gray feathers, reducing the percentage of the population to 2%, which of the following statements accurately describes the gene pool?

Ⓐ The heterozygote advantage has become a homozygous recessive advantage.

Ⓑ Natural selection will make it easier for predators to prey on the chickens with gray feathers.

Ⓒ The population of chickens is no longer in Hardy-Weinberg equilibrium.

Ⓓ The allele frequency will become 0.98 and 0.02.

10. Natural Selection | **251**

Chapter 10 Summary

- Charles Darwin made the following key observations that led to his theory of natural selection:
  - Each species produces more offspring than can survive.
  - Offspring compete with each other for limited resources.
  - Organisms in every population vary.
  - The offspring with the most favorable traits are most likely to survive and reproduce.

- Evidence for evolution includes:
  - fossils
  - biogeography
  - comparison of developmental embryology
  - comparative anatomy, including homologous and analogous structures
  - molecular biology (sequences of genes are conserved across many types of species)

- Members of a species are defined by the ability to reproduce fertile offspring.

- Evolution and speciation occur when environmental pressures favor traits that permit survival and reproduction of selected individuals in a varied population. The favored traits may evolve to cause divergent or convergent evolution.

- Adaptations are random variations in traits that end up being selected for until they become prominent in the population.

- Genetic drift occurs when a population's traits change due to random events rather than natural selection.

- Selection can be stabilizing if an average phenotype is preferred, disruptive if the extremes are preferred, and directional if one extreme is preferred.

- Hardy-Weinberg equations can be used to determine genetic variation within a population. These equations are as follows:
  - $p + q = 1$ Frequency of the dominant (p) and recessive (q) alleles
  - $p^2 + 2pq + q^2 = 1$ Frequency of the homozygous dominants (p^2), heterozygotes ($2pq$), and homozygous recessives (q^2)

- Hardy-Weinberg equations describe a population that is not evolving but is instead said to be in genetic equilibrium. Such a population will be so only if it is not violating any of the following five conditions (almost all populations are evolving due to a violation of one or more of these factors):
    - large population
    - no mutations
    - no immigration or emigration
    - random mating
    - no natural selection

Chapter 11
Ecology

In the preceding chapters, we looked at individual organisms and the ways they solve life's many problems: acquiring nutrition, reproducing, and so on. Now let's turn to how organisms deal with their environments.

INTERACTING WITH THE ENVIRONMENT

Organisms are constantly interacting with their environments. As the environmental stimuli change, so must the behavior and physiology of the organism. This is important for gathering the most energy while using the least energy and reproducing as efficiently as possible.

A general rule is that larger animals have a slower metabolic rate per unit mass than smaller animals. In other words, if you took the same amount of tissue from a small animal and a large animal, the tissue from the small animal would have a higher rate of metabolism.

When animals have excess energy, they grow and store energy. When they are deficient, then they use up their stores and lose mass and eventually die.

Body temperature is one way that animals must respond to their environment. **Endotherms** are animals that generate their own body heat through metabolism. **Ectotherms** lack an internal mechanism to control body temperature. Instead, they regulate their temperature behaviorally by seeking out cool or warm places. One example is a lizard on a warm rock.

BEHAVIOR

Some animals behave in a programmed way to specific stimuli, while others behave according to some type of learning. We humans can do both. For the AP Biology Exam, you'll have to know a little about these different types of **behavior**—that is, how organisms cope with their environments.

Instinct

Innate Selection
Natural selection favors certain behaviors that increase the survival of offspring. Some of these behaviors are instincts. Many learned behaviors are also essential for survival, indicating that the capability to learn might, in fact, be a trait that is naturally selected for.

Instinct is an inborn, unlearned behavior. Sometimes the instinctive behavior is triggered by environmental signals called releasers. The releaser is usually a small part of the environment that is perceived. For example, when a male European robin sees another male robin, the sight of a tuft of red feathers on the male is a releaser that triggers fighting behavior. In fact, because instinct underlies all other behavior, it can be thought of as the circuitry that guides behavior.

For example, hive insects, such as bees and termites, never learn their roles; they are born knowing them. On the basis of this inborn knowledge, or instinct, they carry out all their other behaviors: a worker carries out "worker tasks," a drone "drone tasks," and a queen "queen tasks." Another example of instinct is the "dance" of the honey bee, which is used to communicate the location of food to other members of the beehive.

There are other types of instinct that last for only a part of an animal's life and are gradually replaced by "learned" behavior. For example, human infants are born with an ability to suck from a nipple. If it were not for this instinctual behavior, the infant would starve. Ultimately, however, the infant will move beyond this instinct and learn to feed itself. What exactly, then, is instinct?

For our purposes:

> Instinct is the inherited "circuitry" that directs and guides behavior.

Learning

Another form of behavior is **learning**. Learning refers to a change in a behavior brought about by an experience (which is what you're doing this very moment). Animals learn in a number of ways.

Imprinting

Have you ever seen a group of goslings waddling along after their mother? How is it that they recognize her? Well, the mother arrives and gives out a call that the goslings recognize. The goslings, hearing the call, know that this is their mother, and follow her around until they are big enough to head out on their own.

Now imagine the same goslings, newly hatched. If the mother is absent, they will accept the first moving object they see as their mother. This process is known as **imprinting**.

Animals undergo imprinting within a few days after birth in order to recognize members of their own species. While there are different types of imprinting, including parent, sexual, and song imprinting, they all occur during a **critical period**—a window of time when the animal is sensitive to certain aspects of the environment.

Remember:

> Imprinting is a form of learning that occurs during a brief period of time, usually early in an organism's life.

Habituation

Habituation is another form of learning. It occurs when an animal learns not to respond to a stimulus. For example, if an animal encounters a stimulus over and over again without any consequences, the response to it will gradually lessen and may altogether disappear. Or, if a certain hallway always smells weird, you will notice at first but eventually learn not to notice it. The basic response (smell) has not been lost, merely modified by learning. If you walk down a new hallway and it has the same weird smell, you will notice it.

Internal Clocks: The Circadian Rhythm

There are other instinctual behaviors that occur in both animals and plants. One such behavior deals with time. Have you ever wondered how roosters always know when to start crowing? The first thought that comes to mind is that they've caught a glimpse of the sun. Yet many crow even before the sun has risen.

> **The Science Behind Your Jet Lag**
>
> If you've ever flown overseas, you know all about circadian rhythms. They're the cause of jet lag. Our bodies tell us it's one time, while our watches tell us it's another. The sun may be up, but our body's internal clock is crying, "Sleep!" This sense of time is purely instinctual: you don't need to know how to tell time in order to feel jet lag.

Roosters do have internal alarm clocks. Plants have them as well. These internal clocks, or cycles, are known as **circadian rhythms**. Circadian sounds like the word circle, just like these rhythms have a circular nature.

Watch out though: seasonal changes, like the loss of leaves by deciduous trees or the hibernation of mammals, are not examples of circadian rhythms. *Circadian* refers only to daily rhythms. Need a mnemonic? Just think how bad your jet lag would be after a trip around the world. In other words: circling the globe messes up your circadian rhythm.

HOW ANIMALS COMMUNICATE

A big part of behavior is communication. This is important between members of the same species to coordinate group behavior. Animals use visual, auditory, chemical, tactile, and electrical signals to indicate dominance, find food, establish territory, and reproduce successfully. Communication is also important between members of different species that are engaged in cooperative behaviors.

Some animals use signals as a way of communicating with members of their species. These signals, which can be chemical, visual, electrical, or tactile, are often used to influence mating and social behavior.

Chemical signals are one of the most common forms of communication among animals. **Pheromones**, for instance, are chemical signals between members of the same species that stimulate olfactory receptors and ultimately affect behavior. For example, when female insects give off their pheromones, they attract males from great distances.

Visual signals also play an important role in the behavior observed among members of a species. For example, fireflies produce pulsed flashes that can be seen by other fireflies far away. The flashes are sexual displays that help male and female fireflies identify and locate each other in the dark.

Other animals use electrical channels to communicate. Some fish generate and receive weak electrical fields. Others use tactile signals and have mechanoreceptors in their skin to detect prey. For instance, cave-dwelling fishes use mechanoreceptors in their skin for communicating with other members and detecting prey.

Social Behavior

Many animals are highly social species, and they interact with each other in complex ways. Social behaviors can help members of the species survive and reproduce more successfully. Several behavioral patterns for animal societies are summarized below:

- **Agonistic behavior** is aggressive behavior that occurs as a result of competition for food or other resources. Animals will show aggression toward other members that tend to use the same resources. An example of agonistic behavior is fighting among competitors.

- **Dominance hierarchies** (pecking orders) occur when members in a group have established which members are the most dominant. The more dominant male will often become the leader of the group and will usually have the best pickings of the food and females in the group. Once the dominance hierarchy is established, competition and tension within the group is reduced.

- **Territoriality** is a common behavior when food and nesting sites are in short supply. Usually, the male of the species will establish and defend his territory (called a home range) within a group in order to protect important resources. This behavior is typically found among birds.

- **Altruistic behavior** is defined as unselfish behavior that benefits another organism in the group at the individual's expense because it advances the genes of the group. For example, when ground squirrels give warning calls to alert other squirrels of the presence of a predator, the calling squirrel puts itself at risk of being found by the predator.

Symbiotic Relationships

Many organisms that coexist exhibit some type of **symbiotic relationship.** These include remoras, or "sucker fish," which attach themselves to the backs of sharks, and lichen—the fuzzy, mold-like stuff that grows on rocks. Lichen appears to be one organism, when in fact it is two organisms—a fungus and a photosynthetic organism such as an alga or cyanobacterium—living in a complex symbiotic relationship.

Overall, there are three basic types of symbiotic relationships:

- **Mutualism**—in which both organisms win (for example, the lichen components)
- **Commensalism**—in which one organism lives off another with no harm to the host organism (for example, the remora)
- **Parasitism**—in which the organism actually harms its host

There are many special types of relationships between multicellular organisms and unicellular organisms. One important example is bacteria living inside the guts of many mammals. We have a mutualistic relationship. Humans and other mammals provide the bacteria a nice home with nutrients, and they help us in many ways by breaking things down and helping us make things. It is also good that we are filled with "good" bacteria because otherwise we might be susceptible to dangerous pathogenic bacteria.

PLANT BEHAVIOR

Plants have also evolved specific ways to respond to their environment. The plant behaviors covered on the AP Biology Exam are photoperiodism and tropisms. Plants flower in response to changes in the amount of daylight and darkness they receive. This is called **photoperiodism**. Although you'd think that plants bloom based on the amount of sunlight they receive, they actually flower according to the amount of uninterrupted darkness.

Tropisms

Plants need light. Notice that all the plants in your house tip toward the windows. This movement of plants toward the light is known as phototropism. As you know, plants generally grow up and down: the branches grow upward, while the roots grow downward into the soil, seeking water. This tendency to grow toward or away from the earth is called gravitropism. All of these tropisms are examples of behavior in plants.

A **tropism** is a turning in response to a stimulus.

> Cyanobacteria are bacteria that can perform photosynthesis. They contain chlorophyll pigments but do not contain chloroplast organelles. Algae are eukaryotic protists that can perform photosynthesis.

> Most interactions are performed to gain energy. This can be by directly gaining resources or by doing something else more efficiently so that more energy is available to the organism.

There are three basic tropisms in plants. You can remember them because their prefixes represent the stimuli to which plants react:

- **Phototropism** refers to the way plants respond to sunlight—for example, bending toward light.
- **Gravitropism** refers to the way plants respond to gravity. Stems exhibit negative gravitropism (they grow up, away from the pull of gravity), whereas roots exhibit positive gravitropism (they grow downward into the earth).
- **Thigmotropism** refers to the way plants respond to touch. For example, ivy grows around a post or trellis.

These responses are initiated by hormones. The major plant hormones you need to know belong to a class called **auxins**. Auxins serve many functions in plants. They can promote growth on one side of the plant. For example, in phototropism, the side of the plant that faces away from the sunlight grows faster, thanks to the plant's auxins, making the plant bend toward the light.

Generally speaking, auxins are in the tip of the plant, because this is where most growth occurs. Auxins are also involved in cell elongation and fruit development.

Other plant hormones that regulate the growth and development of plants are **gibberellins, cytokinins, ethylenes,** and **abscisic acid**. Here's a summary of the functions of plant hormones:

FUNCTIONS OF PLANT HORMONES	
Hormone	**Function**
Gibberellins	Promote stem elongation, especially in dwarf plants
Cytokinins	Promote cell division and differentiation
Ethylene	Induces leaf abscission and promotes fruit ripening
Abscisic acid	Inhibits leaf abscission and promotes bud and seed dormancy
Auxins	Promote plant growth and phototropism

We've already discussed how important cell communication is to organisms, and plants are no exception. These hormones play an important role in allowing plants to grow, develop, survive, and adapt to environmental cues. They act as signaling molecules, bind to receptors, and function using signal transduction. These cell communication pathways are similar to what happens in animal cells.

ECOLOGY

The study of interactions between living things and their environments is known as **ecology**. We've spent most of our time discussing individual organisms. However, in the real world, organisms are in constant interaction with other organisms and the environment. The best way for us to understand the various levels of ecology is to progress from the big picture, the biosphere, down to the smallest ecological unit, the population.

There is a hierarchy within the world of ecology. Each of the following terms represents a different level of ecological interaction:

- **Biosphere**: The entire part of the Earth where living things exist. This includes soil, water, light, and air. In comparison to the overall mass of the Earth, the biosphere is relatively small. If you think of the Earth as a basketball, the biosphere is equivalent to a coat of paint over its surface.
- **Ecosystem**: The interaction of living and nonliving things
- **Community**: A group of populations interacting in the same area
- **Population**: A group of individuals that belong to the same species and that are interbreeding

Biosphere

Biomes
There are several major biomes: tundra, taiga, temperate deciduous forest, grasslands, deserts, and tropical rainforests.

The biosphere can be divided into large regions called **biomes**. Biomes are massive areas that are classified mostly on the basis of their climates and plant life. Because of different climates and terrains on the Earth, the distribution of living organisms varies.

Remember that biomes tend to be arranged along particular latitudes. For instance, if you hiked from Alaska to Kansas, you would pass through the following biomes: tundra, taiga, temperate deciduous forests, and grasslands.

Ecosystem

Ecosystems are self-contained regions that include both living and nonliving factors. For example, a lake, its surrounding forest, the atmosphere above it, and all the organisms that live in or feed off the lake would be considered an ecosystem. Living factors are called **biotic factors** and nonliving factors are called **abiotic factors.** Abiotic factors include water, humidity, temperature, soil/atmosphere composition, light, and radiation. As you probably know, there is an exchange of materials between the components of an ecosystem.

Take a look below at the flow of carbon through a typical ecosystem.

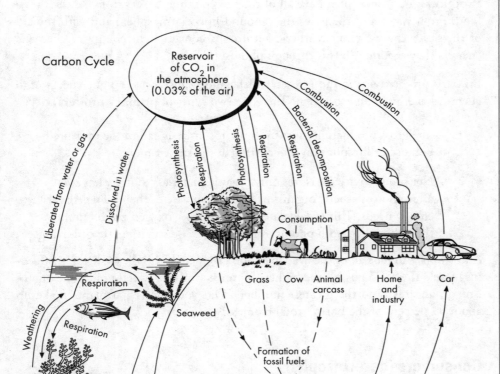

You'll notice how carbon is recycled throughout the ecosystem—this is called the **carbon cycle**. In other words, carbon flows through ecosystems. There are also cycles for nitrogen and water. The interactions between the biotic and the abiotic factors are crucial for balance of the ecosystem.

Community

The next smaller level is the community. A community refers to a group of interacting plants and animals that show some degree of interdependence. For instance, you, your dog, and the fleas on your dog are all members of the same community.

Each organism has its own **niche**—its position or function in a community. When two organisms occupy the same niche, they will compete for the resources within that niche. If a species can occupy an unoccupied niche, it will usually thrive without competition. These connections are shown in the **food chain**. A food chain describes the way different organisms depend on one another for food. There are basically four levels to the food chain: producers, primary consumers, secondary consumers, and tertiary consumers.

Producers (autotrophs)

Producers, or **autotrophs**, have all of the raw building blocks to make their own food. From water and the gases that abound in the atmosphere and with the aid of the Sun's energy, photosynthetic autotrophs convert light energy to chemical energy. They accomplish this through photosynthesis.

Primary productivity is the rate at which autotrophs convert light energy into chemical energy in an ecosystem. There are two types of **primary productivity:**

1. The gross productivity just from photosynthesis cannot be measured because cell respiration is occurring at the same time.

2. Net productivity measures only organic materials that are left over after photosynthetic organisms have taken care of their own cellular energy needs. This is calculated by measuring oxygen production in the light, when both photosynthesis and cell respiration are occurring.

Autotrophs produce all of the available food. They make up the first trophic (feeding) level. They also possess the highest **biomass** (the total weight of all the organisms in an area) and the greatest numbers. Did you know that plants make up about 99 percent of the Earth's total biomass?

> **Chemosynthetic**
> There is another group of autotrophs that are not photosynthetic; chemosynthetics can take up nutrients from their surroundings and make their own energy in the absence of oxygen.

Consumers (heterotrophs)

Consumers, or **heterotrophs**, are forced to find their energy sources in the outside world. Basically, heterotrophs digest the carbohydrates, lipids, proteins, and nucleic acids of their prey into carbon, hydrogen, oxygen, etc., and use these molecules to make organic substances. Recall that organisms digest macromolecules into smaller monomers by utilizing a hydrolysis reaction.

The bottom line is that heterotrophs, or consumers, get their energy from the things they consume.

Primary consumers are organisms that directly feed on producers. A good example is a cow. These organisms are also known as **herbivores**. They make up the second trophic level.

The next level consists of organisms that feed on primary consumers. They are the **secondary consumers**, and they make up the third trophic level. Above these are **tertiary consumers**.

Decomposers

All organisms at some point must finally yield to decomposers. **Decomposers** are the organisms that break down organic matter into simple products. Generally, fungi and bacteria are the decomposers. They serve as the "garbage collectors" in our environment.

> A food chain shows a linear network of links, from producer to apex predator species. A food web consists of many food chains and shows many paths that connect plants and animals.

- Producers make their own food.
- Primary consumers (herbivores) eat producers.
- Secondary consumers (carnivores and omnivores) eat producers and primary consumers.
- Tertiary consumers eat all of the above.
- Decomposers break things down.

Keystone Species

Sometimes one organism is particularly important to an ecosystem. Maybe it is the only producer, or maybe it is the only thing that keeps a particularly deadly predator in check. Important species like this are called **keystone species**. If a keystone species is removed from the ecosystem, the whole balance can be undone very quickly.

The 10% Rule

In a food chain, only about 10 percent of energy is transferred from one level to the next—this is called the **10% rule**. The other 90 percent is used for things like respiration, digestion, reproduction, running away from predators—in other words, it's used to power the organism doing the eating! This energy is eventually changed into heat energy. As a level, producers have the most energy in an ecosystem; the primary consumers have less energy than producers; secondary consumers have less energy than the primary consumers; and tertiary consumers have the least energy of all.

The energy flow, biomass, and numbers of members within an ecosystem can be represented in an **ecological pyramid**. Organisms that are higher up on the pyramid have less biomass and energy, and they are fewer in number.

Not For Individuals
This idea about energy does not apply to individuals. It means the level as a whole has less energy if you add up all the members at that level.

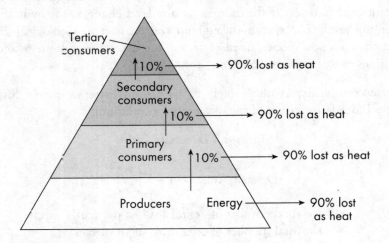

A net gain in energy results in energy storage or growth, and a net loss of energy results in loss of mass and, ultimately, death.

Buildup of Toxins

The downside of food pyramids is that when consumers at the top eat something beneath them, it is like they are eating that thing AND the thing that thing ate AND the thing that thing ate, and so on. So, if there is a toxin in the environment, the consumers at the top are getting the most of it because it becomes more concentrated at each level.

> Toxins in an ecosystem are more concentrated and thus more dangerous for animals further up the pyramid.

This simply means that if a toxin is introduced into an ecosystem, the animals most likely to be affected are those at the top of the pyramid. This occurs because of the increasing concentration of such toxins. The classic example of this phenomenon is DDT, an insecticide initially used to kill mosquitoes. Although the large-scale spraying of DDT resulted in a decrease in the mosquito population, it also wound up killing off ospreys.

Ospreys are aquatic birds of prey whose diet consists primarily of fish. The fish that ospreys consumed had, in turn, been feeding on contaminated insects (**bioaccumulation**). Because fish eat thousands of insects, and ospreys hundreds of fish, the toxins grew increasingly concentrated (**biomagnification**). Though the insecticide seemed harmless enough, it resulted in the near extinction of certain osprey populations. What no one knew was that in sufficient concentrations, DDT weakened the eggshells of ospreys. Consequently, eggs broke before they could hatch, killing the unborn ospreys.

Such environmental tragedies still occur. All ecosystems, small or great, are intricately woven, and any change in one level invariably results in major changes at all the other levels.

Disrupting the Balance

Ecosystems are never exactly stable, existing in a precarious balance. Many things can disrupt the balance. If the biomass at any level changes, this will affect the surrounding levels and spread throughout the ecosystem. Geological changes, weather changes, new species, diseases, lack of resources, new resources, and many other things can disrupt the balance.

A diverse community is more likely to withstand extreme events. Simpson's Diversity Index is a measure of the diversity of a community.

Simpson's Diversity Index

$$\text{Diversity Index} = 1 - \Sigma \left(\frac{n}{N} \right)^2$$

n = the total number of organisms of a particular species
N = the total number of organisms of all species

DDT stands for dichloro-diphenyltrichloroethane. You can see why an abbreviation is better here!

Ecosystem Interconnectedness
Because all levels of an ecosystem are closely connected, a disruption in one level due to environmental changes, for example, affects all other aspects of that ecosystem.

Invasive species are particularly dangerous since they can severely disrupt the balance of the ecosystem. This is often because they have no natural predator or competition and soon expand and take resources from other species.

POPULATION ECOLOGY

Population ecology is the study of how populations change. Whether these changes are long-term or short-term, predictable or unpredictable, we're talking about the growth and distribution patterns of a population.

When studying a population, you need to examine four things: the size (the total number of individuals), the density (the number of individuals per area), the distribution patterns (how individuals in a population are spread out), and the age structure.

For example, the graph below shows there is a high death rate among the young of oysters, but those that survive do well. On the other hand, there is a low death rate among the young of humans, but, after age 60, the death rate is high.

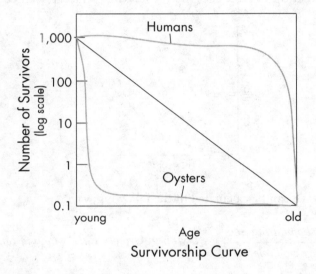

Survivorship Curve

Population growth can be represented as the number of births minus the number of deaths divided by the size of the population.

$$r = \frac{\text{births} - \text{deaths}}{N}$$

(r is the reproductive rate, and N is the population size.)

Population growth can also be calculated in the following way:

$$\frac{\text{Change in Population Size}}{\text{Change in Time}} = \text{Birth Rate} - \text{Death Rate}$$

Each population has a **carrying capacity**—the maximum number of individuals of a species that a habitat can support. Most populations, however, don't reach their carrying capacity because they're exposed to limiting factors.

One important factor is **population density**. The factors that limit a population are either density-independent or density-dependent. **Density-independent factors** affect the population, regardless of the density of the population. Some examples are severe storms and extreme climates. On the other hand, **density-dependent**

Same Equation, Different Look
On the AP Exam, you may see this formula expressed as

$dN/dt = B - D$.
We've just translated the variables to make it easier to interpret. Make sure that you are similarly prepared to parse any formula you're given.

factors are those with effects that depend on population density. Resource depletion, competition, and predation are all examples of density-dependent factors. In fact, these effects become even more intense as the population density increases.

Exponential Growth

The growth rates of populations also vary greatly. There are two types of growth: exponential growth and logistic growth. Equations for both are included on the AP Biology Equations and Formulas sheet. **Exponential growth** occurs when a population is in an ideal environment. Growth is unrestricted because there are lots of resources, space, and no disease or predation. Here's an example of exponential growth. Notice that the curve arches sharply upward—the exponential increase.

Exponential growth can be calculated in the following way:

$$\frac{\text{Change in Population Size}}{\text{Change in Time}} = \text{Maximum Growth Rate} \times \text{Population Size}$$

Exponential growth occurs very quickly, resulting in a J-shaped curve. A good example of exponential growth is the initial growth of bacteria in a culture. There's plenty of room and food, so they multiply rapidly.

However, as the bacterial population increases, the individual bacteria begin to compete with each other for resources. When the resources become limited, the population reaches its carrying capacity, and the curve levels off.

The More Things Change, The More They Stay the Same
As with the equation for population growth, this may be presented to you on the test as $\frac{dN}{dt} = r_{max}N$. Be familiar with what each variable represents.

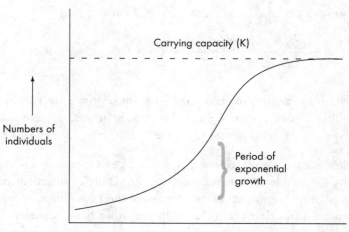

The population becomes restricted in size because of limited resources. This is referred to as **logistic growth**. Notice from the graph on the previous page that the growth forms an S-shaped curve.

This kind of logistic growth can be calculated with the following equation:

$$\frac{Change\ in\ Population\ Size}{Change\ in\ Time} = maximum\ growth\ rate \times Population\ Size \times$$

$$\left(\frac{(Carrying\ Capacity - Population\ Size)}{Carrying\ Capacity}\right)$$

These growth patterns are associated with two kinds of life-history strategies: r-selected species and k-selected species.

Those who thrive in areas that are barren or uninhabited are known as **r-strategists.** Once they colonize an area, they reproduce as quickly as possible. Why? They know they've got to multiply before competitors arrive on the scene! The best way to ensure their survival is to produce lots of offspring. Typical examples are common weeds, dandelions, and bacteria.

At the other end of the spectrum are the **k-strategists**. These organisms are best suited for survival in stable environments. They tend to be large animals, such as elephants, with long lifespans. Unlike r-strategists, they produce few offspring. Given their size, k-strategists usually don't have to contend with competition from other organisms.

> **One More Equation for the Road**
> On the AP Exam, you may see the logistic growth equation written as
> $$\frac{dN}{dt} = R_{max} N \left(\frac{K-N}{K}\right).$$
> Essentially it's the same as the exponential equation except when the population size approaches the carrying capacity, the growth rate approaches zero.

ECOLOGICAL SUCCESSION

Communities of organisms don't just spring up on their own; they develop gradually over time. **Ecological succession** refers to the predictable procession of plant communities over a relatively short period of time (decades or centuries).

Centuries may not seem like a short time to us, but if you consider the enormous stretches of time over which evolution occurs, hundreds of thousands or even millions of years, you'll see that it is pretty short.

The process of ecological succession in which no previous organisms have existed is called **primary succession**.

How does a new habitat full of bare rocks eventually turn into a forest? The first stage of the job usually falls to a community of lichens. Lichens are hardy organisms. They can invade an area, land on bare rocks, and erode the rock surface, over time turning it into soil. Lichens are considered **pioneer organisms**.

Once lichens have made an area more habitable, they've set the stage for other organisms to settle in. Communities establish themselves in an orderly fashion. Lichens are replaced by mosses and ferns, which in turn are replaced by tough grasses, then low shrubs, then evergreen trees, and, finally, deciduous trees. Why are lichens replaced? Because they can't compete with the new plants for sunlight and minerals.

The entire sequence is called a **sere**. The final community is called the **climax community**. The climax community is the most stable. In our example, deciduous trees are part of the climax community.

Now what happens when a forest is devastated by fire? The same principles apply, but the events occur much more rapidly. The only exception is that the first invaders are usually not lichens, but grasses, shrubs, saplings, and weeds. When a new community develops where another community has been destroyed or disrupted, this event is called **secondary succession**. The existing community can be cleared by events such as fire, tornado, or human impact. These impacts leave soil intact, explaining how development occurs more rapidly.

HUMAN IMPACT ON THE ENVIRONMENT

Unfortunately, humans have disturbed the existing ecological balance, and the results are far-reaching. Soils have been eroded and various forms of pollution have increased. Potential consequences on the environment are summarized below.

- **Greenhouse effect:** Atmospheric concentrations of carbon dioxide have increased by burning fossil fuels, and deforestation has contributed to warming of Earth. Higher temperatures may cause polar ice caps to melt and flooding to occur. Other potential effects of global warming include changes in precipitation patterns, changes in plant and animal populations, and detrimental changes in agriculture.

- **Ozone depletion:** Pollution has also led to depletion of the atmospheric ozone layer by such chemicals as chlorofluorocarbons (CFCs), which are used in aerosol cans. Ozone (O_3) forms when UV radiation reacts with O_2. Ozone protects Earth's surface from excessive ultraviolet radiation. Its loss could have major genetic effects and could increase incidence of cancer.

- **Acid rain:** Burning fossil fuels produces pollutants such as sulfur dioxide and nitrogen dioxide. When these compounds react with droplets of atmospheric water in clouds, they form sulfuric and nitric acids, respectively. The rain that falls from these clouds is weakly acidic and is called acid rain. Acid rain lowers the pH of aquatic ecosystems and soil, which damages water systems, plants, and soil. For example, change in soil pH causes calcium and other nutrients to leach out, which damages plant roots and stunts their growth. Furthermore, useful microorganisms that release nutrients from decaying organic matter into the soil are also killed, resulting in fewer nutrients being available for the plants. Low pH also kills fish, especially those that have just hatched.

- **Desertification:** When land is overgrazed by animals, it turns grasslands into deserts and reduces available habitats for organisms.

- **Deforestation:** When forests are cleared (especially by the slash and burn method), erosion, floods, and changes in weather patterns can occur.

- **Pollution:** Another environmental concern is toxic chemicals in our environment. One example is sulfur dioxide, a chemical released when fossil fuels are burned in industrial processes. Sulfur dioxide can cause respiratory issues in living organisms, damage trees and other plants, and cause acid rain. The danger with toxins, such as sulfur dioxide, is that as each trophic level obtains sulfur dioxide, the toxin becomes more concentrated by the process of biomagnification.

- **Reduction in biodiversity:** As different habitats have been destroyed, many plants and animals have become extinct. Some of these plants could have provided us with medicines and products that may have been beneficial.

- **Introduction and spread of disease:** Humans travel and disrupt habitats and bring diseases with them. These diseases can devastate all types of living things and act as immediate selective pressures, leaving the landscape of an ecosystem forever changed.

> Acid rain, deforestation, pollution, soil erosion, and depletion of the ozone layer are just a few of the problems plaguing the environment today.

CHAPTER 11 KEY TERMS

endotherms
ectotherms
behavior
instinct
learning
imprinting
critical period
habituation
circadian rhythms
pheromones
agonistic behavior
dominance hierarchy
territoriality
altruistic behavior
symbiotic relationship
mutualism
commensalism
parasitism
photoperiodism
tropism
phototropism
gravitropism
thigmotropism
auxins
gibberellins
cytokinins
ethylenes
abscisic acid
ecology
biosphere
ecosystem
community
population
biomes
biotic factors
abiotic factors
carbon cycle
niche

food chain
producers (autotrophs)
primary productivity
biomass
consumers (heterotrophs)
primary consumers
herbivores
secondary consumers
tertiary consumers
decomposer
keystone species
10% rule
ecological pyramid
bioaccumulation
biomagnification
Simpson's Diversity Index
population growth
carrying capacity
population density
density-independent factors
density-dependent factors
exponential growth
logistic growth
r-strategists
k-strategists
ecological succession
primary succession
pioneer organisms
sere
climax community
secondary succession
greenhouse effect
ozone depletion
acid rain
desertification
deforestation
pollution

Chapter 11 Drill

Answers and explanations can be found in Chapter 15.

1 | Mark for Review

Turtles lay many eggs in one nest, but only a small number of baby turtles actually hatch and successfully make it to the ocean to survive to adulthood. Which of the following best describes the turtles' survivorship curve?

(A) The curve will increase at a constant positive slope, then level off.

(B) The curve will stay constant then sharply decrease.

(C) The curve will decrease at a constant negative slope then level off.

(D) The curve will decrease at a constant negative slope.

2 | Mark for Review

A community of species in a forest environment is studied. The community's rate of cellular respiration is higher than the rate of photosynthesis. Which of the following explains this scenario?

(A) The laws of thermodynamics are not applicable to this scenario.

(B) The community has reached a climax.

(C) The total biomass of autotrophs in the community is decreasing.

(D) The total biomass in the community is decreasing.

3 | Mark for Review

When a journalist travels south to Panama (a tropical ecosystem) from Canada (an Arctic ecosystem), she passes through tundra, coniferous forest, tropical rain forest, and temperate deciduous forest ecosystems. What causes the significant change in types and quantity of vegetation that she observes through her travels?

(A) Wildfires increase as the journalist travels south.

(B) Mean annual temperature and precipitation increase as the journalist travels south.

(C) Invasive species increase as the journalist travels south.

(D) Primary and secondary succession increase as the journalist travels south.

4 | Mark for Review

Which of the following adaptations helps species obtain energy and nutrients from their surroundings?

(A) The cold-blooded mechanism of a reptile

(B) The claws and teeth of a polar bear

(C) The hibernation abilities of a brown bear

(D) The white fur color of an Arctic rabbit

11. Ecology | **273**

AP Biology Premium Prep

---

5 🔖 Mark for Review

Oahu and Maui are two of the islands that make up the state of Hawaii. Maui is approximately 20% larger than Oahu. Which of the following statements is correct?

(A) Oahu has more competition between individuals than Maui.

(B) Oahu can support larger populations of organisms than Maui can.

(C) Oahu is likely to have a wider range of habitats than Maui.

(D) Oahu is likely to have more individuals than Maui.

---

6 🔖 Mark for Review

Most biomass and energy pyramids representing different ecosystems have a broad base and a narrow point at the top. Which of the following reasons explains this trend?

(A) Matter gets used up at each trophic level and cannot be recycled.

(B) Apex predators are limited by the quantity of toxic substances present in the environment.

(C) Energy is lost from the ecosystem at each trophic level as it is emitted as heat or used for the organism to perform bodily functions.

(D) Apex predators require less energy than primary consumers and producers.

---

7 🔖 Mark for Review

A botanist is comparing two forest communities. Community 1 has equal percentages of Tree Species W, X, Y, and Z. Community 2 is made up of predominantly Tree Species W with approximately 15% of the community composed of Tree Species X, Y, and Z. Which of the following descriptions accurately compares the two communities?

(A) Community 1 has a greater amount of species diversity than community 2.

(B) Community 1 has a greater amount of species richness than community 2.

(C) Community 2 has a greater amount of species diversity than community 1.

(D) Community 2 has a greater amount of species richness than community 1.

---

8 🔖 Mark for Review

In terms of the nitrogen cycle, where does the nitrogen used by primary and secondary consumers come from?

(A) It is inhaled in the form of nitrogen (N_2) gas through cellular respiration.

(B) It is consumed in water in the form of ammonium (NH_4^+) found in fertilizers.

(C) It is consumed in proteins found in animals and plants.

(D) It is consumed in cooked meats in the form of nitrate (NO_3^-).

AP Biology Premium Prep

9 ☐ Mark for Review

A researcher is exploring whether or not bird migration is driven by genetics. Which of the following methods would allow the researcher to answer this question?

(A) The researcher should capture wild birds and determine how long it takes for them to attempt to migrate.

(B) The researcher should compare the migratory habits of different species of birds.

(C) The researcher should compare the migratory habits of birds at different ages.

(D) The researcher should separate offspring from parents and other birds and observe the offspring for migratory behaviors.

10 ☐ Mark for Review

In a pond ecosystem, a disease causes the number of lily pads living on the surface of the pond to decrease. Which of the following effects would likely occur as a result of the reduced number of lily pads?

(A) There might be an increase in the number of cattails living in the pond ecosystem because the competition between the cattails and lily pads will be reduced.

(B) There might be a decrease in the number of beavers living in the pond as beavers feed on lily pads as their primary food source.

(C) There might be an increase in the number of coyotes as coyotes feed on beavers living in the area.

(D) The water quality might decrease as the lily pads help purify the water.

11. Ecology | **275**

Chapter 11 Summary

Behavior

- Behavior is an organism's response to the environment. Behavior can be instinctual (inborn), learned, or social.
  - Types of social behaviors include agonistic behavior, dominance hierarchies, territoriality, and altruistic behavior.

- There are also plant-specific behaviors known as tropisms. The three basic tropisms are phototropism, gravitropism, and thigmotropism.

Ecology

- There are several major biomes (tundra, taiga, temperate deciduous forest, grasslands, deserts, tropical rainforests) that make up the biosphere. Each biosphere contains ecosystems.

- Within an ecosystem are communities, which consist of organisms fulfilling one of three main roles:
  - Producers, or autotrophs, convert light energy to chemical energy via photosynthesis.
  - Consumers, or heterotrophs, acquire energy from the things they consume. Their digestion of carbohydrates produces carbon, hydrogen, and oxygen, which are then used to make organic substances.
  - Decomposers form fossil fuels from the detritus of other organisms in the ecosystem.
  - The 10% rule says that only 10% of the energy consumed from one level will be retained by the higher level that consumed it. The other energy will be spent to perform normal daily activities.

- The smallest unit of ecology is the population. The growth of a population can be found with the equation $(r) = (births - deaths) / N$.

- The carrying capacity is the maximum number of individuals that can be supported by a habitat. Most populations do not reach carrying capacity due to factors such as population density (density-independent factors and density-dependent factors).

- Exponential growth (J-shaped curve) occurs when a population is in an ideal environment. Logistic growth (S-shaped curve) of a population occurs when there are limited resources in an environment.

- Organisms are generally either r-strategists or k-strategists. The r-strategists ensure their survival by producing lots of offspring. The k-strategists, which are usually large animals, produce few offspring but have longer life spans and less competition from other organisms for resources.

- Succession describes the way in which ecosystems recover after a disturbance in terms of pioneers, sere, climax community, and secondary succession.

- Human impact on the planet includes the following issues:
  - greenhouse effect
  - ozone depletion
  - acid rain
  - desertification
  - deforestation
  - pollution
  - reduction in biodiversity
  - introduction and spread of disease

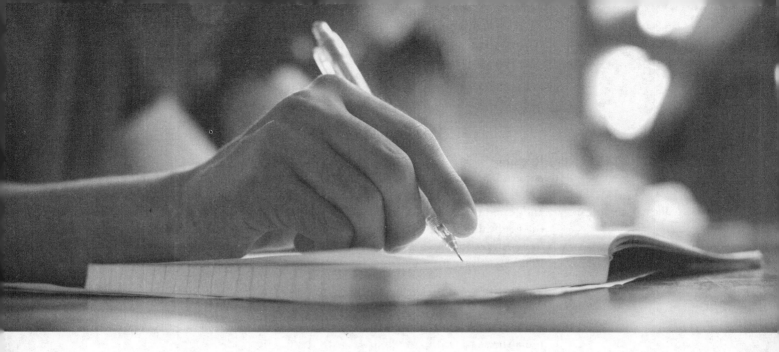

Chapter 12
Quantitative Skills and Biostatistics

AP Biology Premium Prep

Questions related to quantitative skills and biostatistics pop up all over the place. Even if you think you are a whiz, check out the formula pages near the back of this book. Expect at least a couple questions using formulas from these sheets.

It is often easy to see patterns in data, but it is not always easy to determine if a pattern is valid or significant. Quantitative data analysis is the first step in figuring this out. In this chapter, we will review how data can be summarized, presented, and tested for validity. This depends on what kind of question was being asked at the beginning of the experiment. When you're reading this chapter, pay attention to which techniques are used when. On the AP Biology Exam, you should be able to justify which techniques are best.

Don't forget to check out the formula sheet!

This list is a good reference for the types of quantitative questions you should be able to tackle. We will not walk through all of these in detail, but you will find some examples in the end-of-chapter drills and in the practice tests. Use the formula sheets to familiarize yourself with these concepts.

> **Must-Know Types of Quantitative Problems**
>
> ☐ Hardy-Weinberg Equilibrium
>
> ☐ Water Potential/Osmosis
>
> ☐ Energy Pyramids/Biomass
>
> ☐ Chi-squared Analysis
>
> ☐ Gene Linkage Analysis
>
> ☐ Inheritance I Probability
>
> ☐ Reading from Graphs
>
> ☐ Predicting from Graphs
>
> ☐ Population Growth

SUMMARIZING AND PRESENTING DATA

Instead of presenting raw data, scientists use **descriptive statistics** and graphs to summarize large datasets, present patterns in data, and communicate results. Descriptive statistics summarize and show variation in the data. There are many types of descriptive statistics, including measures of central tendency (such as mean, median, and mode) and measures of variability (such as standard deviation, standard error, and range). The AP Biology Equations and Formulas sheet contains definitions for mean, median, mode, and range. It also includes equations for mean, standard deviation, and standard error. That said, you will not need to calculate standard deviation or standard error. Instead, focus on understanding how these values are used.

280 | For more free content, visit PrincetonReview.com

Descriptive statistics are used to summarize the data collected from samples, but may also describe the entire population you're trying to study. Experiments are designed to include a **sample**, which is a subset of the **population** being studied. The best experiments use **random sampling**, which makes sure there is no bias in picking which individuals from the population will be included in the sample. It is important that the sample is big enough that the data from the experiment is a good representation of what would happen if the whole population were measured.

Graphs are visual representations of data and are often used to reveal trends that might not be obvious by looking at a table of numbers. There are six types of graphs you should be familiar with:

1. Bar graph
2. Pie graph
3. Histogram
4. Line graph
5. Box-and-whisker plot
6. Scatterplot

Each of these is described more in the sections that follow. Pay attention to which graph is used when. On the AP Biology Exam, you may need to decide which graph is best and then create it. You may also have to analyze graphs given to you on the exam.

A good graph must include the following things:

- a good title that describes what the experiment was and what was measured
- axes labeled with numbers, labels and units, and index marks
- a frame or perimeter
- data points that are clearly marked (e.g., easily identifiable lines or bars)

TYPES OF DATA

Data can be **quantitative** (based on numbers or amounts that can be measured or counted) or **qualitative** (data that is descriptive, subjective, or difficult to measure). For example, behavioral observations are qualitative. Most data on the AP Biology Exam will be quantitative, with either counts or measurements.

Count Data

Count data are generated by counting the number of items that fit into a category. For example, you could count the number of organisms with a particular phenotype or the number of animals in one habitat versus another. Count data also include data that is collected as percentages or the results of a genetic cross. This type of data is usually summarized in a **bar graph** or a **pie graph**. For example, suppose a mixed population of *E. coli* were plated on growth plates. Some cells contained a β-galactosidase marker and grew in blue colonies, while others did not contain the marker and grew in white colonies. The colonies were counted after 24 hours of growth.

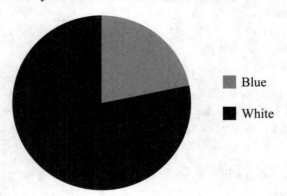

E.coli Colony Growth on Plates after 24 hours

In another example, suppose fly populations were monitored in northern Maine to determine if population size varies over the warm months of the year.

Count data can be analyzed with hypothesis testing, which will be described later in this chapter.

Measurements

Measurements are continuous, meaning there is an infinite number of potential measurements over a given range. Size, height, temperature, weight, and response rate are all measurements. There are two types of measurement data: parametric and nonparametric.

Normal or Parametric Data

Normal, or **parametric**, **data** is measurement data that fits a **normal curve**, or **normal distribution**, usually when a large sample size is used. For example, if you took a large sample of 17-year-olds in America and graphed the frequency of heights, the results will be normally distributed.

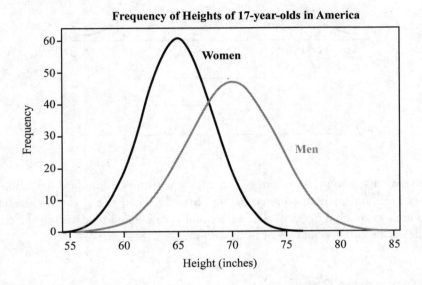

Several descriptive statistics can be used to summarize normal data.

The **sample size** (n) refers to the number of members of the population that are included in the study. Sample size is an important consideration when you're trying to determine how well the data in the study represents a population. Large sample sizes are always better, but for technical reasons, experiments can't be infinitely large.

The **mean** ($\bar{x}$) is the average of the sample, calculated by adding all of the individual values and dividing by the number of values you have. The mean is not necessarily a number provided in the sample. You should be able to recognize what the mean of a given dataset is and be able to calculate the mean using the equation on the AP Biology Equations and Formulas sheet. One limitation of mean is that it is influenced by **outliers**, or numerical observations that are far removed from the rest of the observations.

The **standard deviation** (s) can determine if numbers are packed together or dispersed, because it is a measure of how much each individual number differs from the mean. A low standard deviation means the data points are all similar and close to the mean, while a high standard deviation means the data are more spread out. Here is the relationship between a normal distribution and standard deviation (SD).

This means that about 70 percent of the data is within one standard deviation of the mean in a normally distributed dataset. **Standard error** (SE) can also be used to report how much a given dataset varies and is calculated by dividing the standard deviation by the square root of the sample size.

Nonparametric Data

Nonparametric data often includes large outliers and does not fit a normal distribution. In order to determine if data is parametric or not, you can construct a **histogram**, or **frequency diagram**. These graphs give information on the spread of the data and the central tendencies. Making a histogram is like setting up bins, or intervals with the same range, that cover the entire dataset (the *x*-axis). The measurements in each bin are graphed on the *y*-axis. For example, suppose you counted the number of pine needles on each branch of a *Pinus strobus* tree. A histogram of this data may look like this.

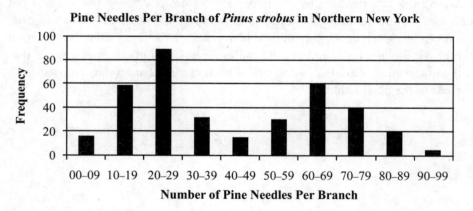

This dataset does not match a normal distribution, meaning it is a nonparametric dataset. Keep in mind that even if a certain dataset is not normal, the population itself could be. There could have been sampling bias or errors in the data collection, which can lead to sample data that doesn't match population data. The best way to fix this is usually to increase the sample size. Histograms differ from bar graphs in that they show ranges instead of just categories.

Nonparametric data requires a different set of descriptive statistics than parametric data.

The **median** is the middle number in a dataset and is determined by putting the numbers in consecutive order and finding the middle number. If there is an odd number of numbers, there will be a single number that is the median. If there is an even number of numbers, the median is determined by averaging the two middle numbers. Therefore, the median is not necessarily one of the numbers in the dataset. The median is useful in gauging the midpoint of the data, but will not necessarily tell you much about outliers.

The **mode** is the most frequently recurring number in the dataset. If there are no numbers that occur more than once, there is no mode. If there are multiple numbers that occur most frequently, each of those numbers is a mode. The mode must be one of the numbers in the sample, and modes are never averaged.

The **range** is the difference between the smallest and largest number in a sample. This value is less useful than standard deviation, because it does not give any information on individual values or the majority of values.

Example 1: Eleven male laboratory mice were weighed at 32 weeks of age. The data collected was: 32 g, 28 g, 29 g, 34 g, 30 g, 28 g, 32 g, 31 g, 30 g, 32 g, 33 g. What is the mean, median, and mode of this dataset?

Solution: Let's start by putting the 11 numbers in the dataset in order, from smallest value to largest value: 28 g, 28 g, 29 g, 30 g, 30 g, 31 g, 32 g, 32 g, 32 g, 33 g, 34 g. The mean is

$$Mean = \frac{28 + 28 + 29 + 30 + 30 + 31 + 32 + 32 + 32 + 33 + 34}{11}$$

$$= \frac{339}{11}$$

$$= 30.8\,g$$

The median is the middle value, or 31 g. The mode is the most frequent number, or 32 g.

Example 2: In the following set of values, what is the range?

Values: –5, 8, 11, –1, 0, 4, 14

Solution: The smallest value in the set above is –5, and the largest is 14. The difference between these two is the range, which is 19.

TYPES OF EXPERIMENTS OR QUESTIONS

You will be tested on scientific experiments and the scientific reasoning behind different aspects of the experiments a LOT. You should be clear with the steps of the scientific method and know how to properly design an experiment and recognize a poorly designed experiment. The following list should remind you:

- **Hypothesis:** A well-thought-out prediction of what the outcome of the experiment will be.
- **Independent Variable:** The factor that you, as the experimenter, will change between the different groups in the experiment. If one plant is grown in pH 5 and the other in pH 7, then pH is the independent variable.
- **Dependent Variable:** The data, or the thing that you measure during the experiment. The height of the plant you are growing might be the dependent variable.
- **Constants (Controlled Variables):** The things that are the same between all your groups. Hint: everything should be constant except the independent variable.
- **Control Groups:** Any group that is needed simply so you can compare your interesting experimental groups to it. These are often the "no treatment" group.
- **Statistical Significance:** The trustworthiness of the results and the certainty you have in your conclusions. This can always be increased by including more individuals in the groups or including more trials.

Most biological experiments do one of three things:

1. look at how something changes over time
2. compare groups of some sort
3. test for an association

How you present and summarize data depends on what kind of experiment you perform.

Time-Course Experiments

Time-course experiments look at how something changes over time. A **line graph** is usually used to present this type of data. Several lines can be plotted on the same graph, but these must be clearly labeled. Also, each dot on a line graph could represent one data point or a mean of values. If mean values are plotted, standard deviation or standard error can also be shown. More generally, line graphs can also be used to compare the way in which a dependent variable (y-axis) changes in relation to an independent variable (x-axis).

When making line graphs, the intervals on the axes must be consistent. Also make sure it is clear whether the data starts at the origin (0, 0) or not. The same guidelines apply to scatterplots, which will be discussed below.

Data points should be connected with a solid line. You can extrapolate (extend) the line past the data points, but then you must use a broken line. Finally, the slope on a line graph tells you the rate of change.

Suppose we want to determine how stress hormone levels change after the end of an important exam. If we measure cortisol and epinephrine in the serum of several people every hour for five hours after the end of an exam, we would have several values for each time point. We could plot the mean of each measurement and also show data variance using standard error bars. In this example, the distance above the data point is the standard error, and the distance below the data point is the standard error. We call this "±SE."

Comparative Experiments

Comparative experiments compare populations, groups, or events. There are several options for graphing data from these experiments:

- **Bar graphs** are helpful to compare categories of data if the data is parametric. It is also a good idea to show variance of the data by including the standard error. This gives information on how different the two means are from each other. Suppose you compare two strains of fission yeast before and after treatment with a drug that inhibits cell wall synthesis. As a measure of viability, you measure how much Myc mRNA was made in each strain with and without treatment. In this case, each bar is a mean of values and the error bars are ±SE, as described above.

- **Box-and-whisker plots** should be used for nonparametric data. These graphs show median (horizontal line), quartiles (top and bottom of the box), and largest and smallest values (ticks at the tops and bottoms of the lines). In other words, the middle 50% of plant weights is represented by the box, with the median shown with the line, and the highest and lowest values are marked with the top and bottom extenders (whiskers). Remember, the median is not the average, so the line doesn't have to be in the middle. Suppose the weights of two types of Arizona plants are measured, to determine which is more plentiful in a given niche:

Grass and Cactus Weights in the Arizona Desert

In order to conclude that the samples or groups being compared are different (or not), hypothesis testing should be performed. This will be discussed later in this chapter.

Association Experiments

Association experiments look for associations between variables. They attempt to determine if two variables are correlated, and additional tests can demonstrate causation. **Scatterplots** are used to present data from association experiments. Each data point is plotted as a dot. Suppose different substrate concentrations are used in an enzymatic reaction (using the enzyme Kozmase III), and the enzyme efficiency is measured (in percent of maximum). The data could be presented like this.

Enzyme Efficiency and Substrate Concentration for Kozmase III at Physiological Temperature and pH

If the relationship looks linear, a linear regression line can be added. There are a few common shapes of scatterplots you should be familiar with:

- A bell-shaped curve is associated with random samples and normal distributions.
- An upward curve (concave in shape) is associated with exponentially increasing functions. A common example of this is the lag phase of bacterial growth in a liquid culture.
- A curve that looks like the sine wave is common when biological rhythms are being studied.

PROBABILITY

You already saw an introduction to probability calculations in Chapter 8 of this book. The **probability** (P) that an event will occur is the number of favorable cases (a) divided by the total number of possible cases (n).

$$P = \frac{a}{n}$$

This can be determined experimentally by observation (such as when population data is being collected) or by the nature of the event. For example, the probability of getting a two when rolling a die is 1/6, since there are six sides on a die.

Example: What is the probability of drawing the nine of hearts from a deck of cards?

Solution: Since there are 52 cards in a deck and this question is asking about one particular card, $P = 1/52$.

Combining Probabilities

Many genetics problems involve several probabilities, all being considered together to answer a larger biological question. There are three rules that are often used.

Product Rule

The **product rule** is used for independent events and is also called the "AND rule." The probability of independent events occurring together is the product of their separate probabilities. This rule is used when the order of the events matters.

Example: What is the probability of drawing a heart and a spade consecutively from a deck of cards if the first draw was replaced before the second draw?

Solution:

$$P(\text{heart}) = \frac{13}{52} = \frac{1}{4}$$

$$P(\text{heart and spade}) = P(\text{heart}) \times P(\text{spade})$$

$$P(\text{spade}) = \frac{13}{52} = \frac{1}{4}$$

$$P(\text{heart and spade}) = \frac{1}{4} \times \frac{1}{4} = \frac{1}{16}$$

Sum Rule

The **sum rule** is used for studying two mutually exclusive events, and can be thought of as the "EITHER" rule. The probability of either of two events occurring is the sum of their individual probabilities.

Example: What is the probability of drawing a diamond or a heart from a deck of cards?

Solution:

$$P(\text{diamond}) = \frac{13}{52} = \frac{1}{4}$$

$$P(\text{heart}) = \frac{13}{52} = \frac{1}{4}$$

$$P(\text{diamond or heart}) = P(\text{diamond}) + P(\text{heart})$$

$$P(\text{diamond or heart}) = \frac{1}{4} + \frac{1}{4} = \frac{2}{4} = \frac{1}{2}$$

If the events you are studying are not mutually exclusive, such as having brown hair and brown eyes, then you could have both occur at the same time. If you are trying to determine the likelihood of EITHER occurring but NOT BOTH, then you must calculate using the sum rule (odds of either occurring) and the product rule (odds of both occurring). Then, take the result using the product rule and subtract it from the result when using the sum rule since you want the odds of either occurring but not both.

HYPOTHESIS TESTING

t-test and *p*-values

Many experiments involve comparing two datasets or two groups, and a **t-test** can be used to calculate whether the means of two groups are significantly different from each other. This test is most often applied to datasets that are normally distributed. A **p-value** equal to or below 0.05 is considered statistically significant in most biology-related fields. *T*-test calculations involve comparing each data point to the group's mean and also take standard deviation and sample size into consideration.

Understanding mathematically how *p*-values are calculated and *t*-tests are performed is not important here. Instead, it's important that you understand how these values are used to interpret data.

AP Biology Premium Prep

Let's work through an example to demonstrate how common statistics are used in biological labs.

A researcher has sections of two different types of skin cancers from human patients. She stains them for the protein CD31, which is a marker of endothelial cells. She then takes digital pictures of the immunofluorescent sections and counts the number of blood vessels present in each picture. She takes five pictures of each slide.

Slide A	Slide B
10	3
8	2
7	4
8	4
11	4

(a)   What is the mean, median, and mode for each dataset?

(b)   The standard error of group A is 0.735, and the standard error for group B is 0.400. What does this tell you about the data?

(c)   What assumptions are made about the data in performing these tests?

(d)   What could the researcher do to increase her confidence in the results?

Solutions:

(a)   The means are

$$Mean_{Group\ A} = \frac{10 + 8 + 7 + 8 + 11}{5} = 8.8$$

$$Mean_{Group\ B} = \frac{3 + 2 + 4 + 4 + 4}{5} = 3.4$$

Remember, the median is the middle number, so it's best to put the data in order first.

Slide A	Slide B
7	2
8	3
8	4
10	4
11	4

The median of group A is 8, and the median of group B is 4.

The mode is the most frequent value. For group A, this is 8. For group B, the mode is 4.

(b) Standard error is the standard deviation divided by the square root of the sample size. The sample size is five for both group A and group B because five pictures were taken for each slide. Because the standard error of group A is larger than that of group B and because the two groups have the same sample size, the standard deviation of group A must also be larger than that of group B. Note that this may or may not be the case if the sample sizes were different. A larger standard deviation means the data is more variable, so you can conclude that the data from group A has a larger spread than the data from group B.

(c) As with most datasets, the researcher is assuming the data fits a normal distribution and that her sample size is large enough to be meaningful.

(d) More data allows for more confident conclusions. The researcher could therefore take more pictures from each slide to increase the sample size.

Chi-Square (χ^2) Tests

A **chi-square test** is a statistical tool used to measure the difference between observed and expected data. For example, suppose you roll a die 120 times. You would expect a 1:1:1:1:1:1 ratio between all the numbers that come up. In other words, you would expect to see each side 20 times. However, because this is an experiment and the sample size is only 120, you will probably see some variation in the data. If you rolled 24 sixes, for example, the disagreement between observed (24) and expected (20) is small, and could have occurred by chance. If you rolled 40 sixes, however, there is a large disagreement from the expected. Maybe it is a weighted die. The region in between these two values is trickier. Where is the line drawn between expected/normal variations and results that are so different that they must tell a different story altogether?

Chi-square tests are one way to make this decision. They start with a **null hypothesis** (H_0), which represents a set numerical outcome that you want to compare your data with. The chi-square test will determine if your actual data is close to the null hypothesis outcome and could have occurred due to a fluke (i.e., rolling a six 24 times) or if it is so different that it is probably not a fluke (i.e., 40 rolled sixes) and the null hypothesis outcome must not correctly describe what is occurring. In the dice case, rolling a six 40 times would cause the null hypothesis of expecting a 1:1:1:1:1:1 to be rejected and the die can be considered to be weighted. In the calculations, expected results if the null hypothesis were true are compared to the actual values, and an χ^2 value is calculated. This value is compared to a **critical value**, and a decision is made to reject or accept the null hypothesis.

For example, suppose data was collected from 100 families with three children each. Fourteen families had three female children, 36 families had two females and a male, 30 families had one female and two males, and 20 families had three male children. We would expect offspring sex to segregate independently. In other words, the sex of a given child shouldn't be affected by the sex of previous children. Let's test whether this is true for this dataset.

Since we expect sex to segregate independently, this is the null hypothesis (H_0). Each family has three children, so there are four possible outcomes for each family:

1. three daughters
2. two daughters and a son
3. one daughter and two sons
4. three sons

Using some advanced math (that you don't need to worry about), the probabilities for each of these is:

1. P(three females) = 12.5%
2. P(two females, one male) = 37.5%
3. P(one female, two males) = 37.5%
4. P(three males) = $(1)(0.50)^3$ = 12.5%

Because there are 100 families, we expect 12.5 to have three females, 37.5 to have two females and a male, 37.5 to have one female and two males, and 12.5 to have three males. This can be calculated by multiplying the probability of each event by the total (100 in this case).

Next, we compare this expected data (E) to what actually happened (the observed data, O). Note that the total numbers for observed and expected columns must be the same (in this case, they each have a total of 100). To calculate an χ^2 value, the formula $(O - E)^2/E$ is calculated for each row of the table, and summed.

Family Type	Observed (O)	Expected (E)	$(O - E)^2/E$
3 females	14	12.5	0.18
2 females 1 male	36	37.5	0.06
1 female 2 males	30	37.5	1.5
3 males	20	12.5	4.5
Total	**100**	**100**	**6.24**

The calculated χ^2 value is 6.24. Next, this value needs to be compared to a critical value (CV). This is obtained from a table like this one.

	Chi-Square Table							
p-value	Degrees of Freedom							
	1	2	3	4	5	6	7	8
0.05	3.84	5.99	7.82	9.49	11.07	12.59	14.07	15.51
0.01	6.64	9.21	11.34	13.28	15.09	16.81	18.48	20.09

This chart is included on the AP Biology Equations and Formulas sheet. In order to use this table, you must know the **degrees of freedom** (DF), which represent the number of independent variables in the data. In most cases, this is the number of possibilities being compared minus one. In this example, DF = 4 − 1, because we are comparing four different family types.

You also need a reference *p*-value. This is the probability of observing a deviation from the expected results due to chance. For most biological tests, a *p*-value of 0.05 is used.

Since we are using DF = 3 and $p = 0.05$, the critical value is 7.82 (according to the chart above). Next, the critical value is compared to the χ^2 value; if $\chi^2 < CV$, you accept H_0. If $\chi^2 > CV$, you reject H_o. We calculated an χ^2 value of 6.24, which is smaller than 7.82. Since $\chi^2 < CV$, we accept H_o and conclude that offspring sex could, indeed, be segregating independently in this dataset.

CHAPTER 12 KEY TERMS

descriptive statistics
sample
population
random sampling
quantitative data
qualitative data
count data
bar graph
pie graph
normal, or parametric, data
normal curve, or normal distribution
sample size
mean
outliers
standard deviation
standard error
nonparametric data
histogram, or frequency diagram
median
mode
range
hypothesis

independent variable
dependent variable
constants (controlled variables)
control groups
statistical significance
time-course experiment
line graph
comparative experiments
bar graph
box-and-whisker plot
association experiments
scatterplot
probability
product rule
sum rule
t-test
p-value
chi-square test
null hypothesis
critical value
degrees of freedom

AP Biology Premium Prep

Chapter 12 Drill

Answers and explanations can be found in Chapter 15.

1 Mark for Review

The height of seven trees was measured before and after a new fertilizer was applied for 6 months. The data is shown in the table below. What is the average change in height, in inches, for all seven trees, rounded to the nearest inch?

Tree	Initial Height (inches)	Final Height (inches)
1	360	372
2	320	344
3	400	410
4	410	425
5	375	402
6	331	351
7	393	405

Ⓐ 7 inches

Ⓑ 17 inches

Ⓒ 20 inches

Ⓓ 120 inches

2 Mark for Review

The size of eight cells is measured in micrometers (μm) and shown in the table below. Is cell A smaller than the median for the data set?

Cell	Size (μm)
A	10.0
B	0.1
C	5.0
D	2.5
E	100.0
F	2.0
G	15.0
H	50.0

Ⓐ Cell A is smaller than the median, which is 48.75.

Ⓑ Cell A is smaller than the median, which is 7.5.

Ⓒ Cell A is not smaller than the median, which is 48.75.

Ⓓ Cell A is not smaller than the median, which is 7.5.

3 Mark for Review

What are the range and mode of the following data set, which represents the number of eggs released by different ocean sunfish during their spawning season (in millions)?

Eggs: 100, 110, 221, 107, 333, 455, 307, 411, 156, 209, 390, 407, 107, 500, 220, 301

Ⓐ The range is 400, and the mode is 107.

Ⓑ The range is 400, and the mode is 127.5.

Ⓒ The range is 201, and the mode is 107.

Ⓓ The range is 201, and the mode is 127.5

12. Quantitative Skills and Biostatistics | **297**

AP Biology Premium Prep

Questions 4 through 5 refer to the following.

A geneticist is studying three traits in canines: coat color, tail length, and ear shape. The gene for a black coat color (B) is dominant over brown (b). The gene for a long tail (L) is dominant over short (l). The gene for floppy ears (F) is dominant over erect (f). Two dogs with the following genotypes are crossed: $BbLlFf \times bbllff$.

---

4 ☐ Mark for Review

What is the probability of the first puppy having the genotype *bbllff*?

Ⓐ 6.25%

Ⓑ 12.5%

Ⓒ 25%

Ⓓ 50%

---

5 ☐ Mark for Review

What is the probability of the first puppy having the genotype *BbLlFf* or *Bbllff*?

Ⓐ 6.25%

Ⓑ 12.5%

Ⓒ 25%

Ⓓ 50%

Questions 6 through 7 refer to the following.

A researcher is studying genetic traits in fruit flies. A mating is set up between two pure breeding fruit flies. One parent has red eyes and long wings, and the other parent has white eyes and short wings. F_1 offspring all have red eyes and long leaves. When F_1 offspring are mated with each other, 500 F_2 fruit flies are analyzed. The data is shown below.

Phenotype	# of F_2 Fruit Flies
Red eyes, long wings	190
Red eyes, short wings	56
White eyes, long wings	54
White eyes, short wings	200
Total	500

---

6 ☐ Mark for Review

Are the genes for eye color and wing length experiencing independent segregation?

Ⓐ Yes; the χ^2 value is smaller than the critical value for a p-value of 0.01, and the degrees of freedom are 1.

Ⓑ Yes; the χ^2 value is smaller than the critical value for a p-value of 0.01, and the degrees of freedom are 3.

Ⓒ No, the χ^2 value is larger than the critical value for a p-value of 0.01, and the degrees of freedom are 1.

Ⓓ No, the χ^2 value is larger than the critical value for a p-value of 0.01, and the degrees of freedom are 3.

---

7 ☐ Mark for Review

Two fruit flies from the F_1 generation are crossed, and the offspring are expected to have a ratio of 3 flies with red eyes to 1 fly with white eyes. When 50 of the offspring are analyzed, it is found that 39 flies have red eyes and 11 flies have white eyes. Does this result reasonably match the expected result for a p-value of 0.01?

Ⓐ Yes; the χ^2 value is smaller than the critical value of 6.63.

Ⓑ No; the χ^2 value is smaller than the critical value of 6.63.

Ⓒ Yes; the χ^2 value is greater than the critical value of 6.63.

Ⓓ No, the χ^2 value is greater than the critical value of 6.63.

Chapter 12 Summary

○ Descriptive statistics (such as mean, standard deviation, standard error, median, mode, range) and graphs are used to summarize data and show patterns and conclusions:
- Normal (parametric) measurement data are summarized by using mean and standard deviation or standard error.
- Parametric data are summarized by using median, mode, and range.
- Histograms can be used to see if a dataset has a standard deviation or not.

○ Count data is summarized in a bar graph or a pie graph.

○ Time course experiments look at how something changes over time; summarize these using a line graph.

○ Experiments that compare groups are summarized in a bar graph (if the dataset has a standard deviation) or a box-and-whisker plot (if it does not).

○ Scatterplots summarize association experiments, and regression lines can be used to determine if the relationship is linear.

○ Probability can be calculated with the sum rule or the product rule.

○ Hypothesis testing is used to determine if two groups are significantly different from each other. They start with a null hypothesis, which is rejected or accepted, depending on how a calculated p-value or chi-square value compares to a standard value.

Chapter 13
Sample Free-Response Questions

AP Biology Premium Prep

SECTION II: FREE RESPONSE

Section II of the AP Biology Exam contains six questions, consisting of two long free-response questions and four short free-response questions. You will have a total of 90 minutes to complete all questions. The directions will look something like this (read carefully!).

Directions: Questions 1 and 2 are long free-response questions that should require about 25 minutes each to answer and are worth 8–10 points each. Questions 3 through 6 are short free-response questions that should require about 10 minutes each to answer. Questions 3 through 6 are worth 4 points each.

Read each question carefully and completely. Write your response in the space provided following each question. Only material written in the space provided will be scored. Answers must be written out in paragraph form. Outlines, bulleted lists, or diagrams alone are not acceptable unless specifically requested.

Remember the Questions

For a full list of the six types of questions and how they're scored, refer to pages 64–66.

Free-Response Tips and Strategies

AP Biology Readers (the people who grade your free-response essays) expect students to interpret data and apply knowledge to new situations in the free-response questions. Some important things to keep in mind when writing your answers to the free-response questions are provided in the following list:

- **Answer the question.** Do not become distracted or waste time writing about tangential topics that you know more about than what is being asked. Focus your response.

- **Think quantitatively.** Use data whenever possible to explain your answer. This is especially necessary when the question features a graph, table, or other type of diagram (which many free-response questions now do).

- **Be specific.** Offer specific detailed examples where appropriate.

- **Use vocabulary correctly and often.** Hone your understanding of the words provided in the Key Terms lists in this book and the various biological ideas and situations to which they can be applied. This will help you use the correct terminology when writing your responses.

- **Remember the Big Ideas.** Know the four big ideas covered in the AP Biology course (page 62), as they are reflected in many free-response questions. Understanding these ideas and being able to articulate them will help you craft your responses.

Okay, now it's time for some practice. Take about 25 minutes and try to write a long free-response to the following question. This is an example of Long Question #2.

302 | For more free content, visit PrincetonReview.com

AP Biology Premium Prep

2 ☐ Mark for Review

The regulation of the expression of gene products is crucial to the development of an embryo and the maintenance of homeostasis.

   A.

       i.   **Describe** differences in maintaining homeostasis in unicellular organisms.

       ii.   **Describe** differences in maintaining homeostasis in multicellular organisms.

The table below illustrates the mRNA levels, total protein levels, and total enzyme activity levels for three different enzymes. These enzymes are encoded by genes A, B, and C after being expressed in both the absence and presence of "Protein X" or "Protein Y."

	Untreated			+ Protein X			+ Protein Y		
	Gene A	Gene B	Gene C	Gene A	Gene B	Gene C	Gene A	Gene B	Gene C
mRNA Levels	1,200	2,000	1,400	2,100	250	1,400	1,200	2,000	1,400
Protein Levels	400	1,100	300	750	20	300	400	1,100	300
Total Enzyme Activity Levels	20,000	32,000	18,000	35,000	5	18,000	20,000	50,000	200
Std Error mRNA	100	132	154	172	23	120	132	186	144
Std Error Protein	43	101	22	37	2	38	20	89	21
Std Error Total Enzyme Activity	1,200	1,500	1,000	2,600	1	900	875	3,200	28
Total Enzyme Activity Standard Deviation	3,000	5,200	3,200	4,000	3	2,000	2,500	800	30

   B.   **Construct** a graph of the total enzyme activity levels for each of the 3 genes. Be sure to show the results in the presence and absence of Protein X and Protein Y.

   C.

       i.   **Describe** the influence of Protein X and Protein Y on the gene expression of Enzyme A. Include at which point in gene expression the regulation likely occurs.

       ii.   **Describe** the influence of Protein X and Protein Y on the gene expression of Enzyme B. Include at which point in gene expression the regulation likely occurs.

   D.   Another scientist repeats the experiment with Protein Y and Enzyme C and adds a known inhibitor to Protein Y to the tube.

       i.   **Predict** the effect on mRNA and protein levels as a result.

       ii.   **Predict** the effect on total enzyme activity levels as a result.

13. Sample Free-Response Questions | **303**

AP Biology Premium Prep

SCORING GUIDELINES

To help you grade this sample free response, we've put together a checklist that you can use to calculate the number of points that should be assigned to each part of this question. We'll first explain the important points on the checklist and then give you sample student responses to show you how test reviewers would evaluate them.

Free-Response Checklist

A.

   i. **Describe** differences in maintaining homeostasis in unicellular organisms.

      • 1 point for describing homeostasis in unicellular organism

   ii. **Describe** differences in maintaining homeostasis in multicellular organisms.

      • 1 point for describing homeostasis in multicellular organisms

B. **Construct** a graph of the total enzyme activity levels for each of the 3 genes. Be sure to show the results in the presence and absence of Protein X and Protein Y.

    • 1 point for including proper data selection/data placement
    • 1 point for proper axes
    • 1 point for proper error bars
    • 1 point for proper labels

C.

   i. **Describe** the influence of Protein X and Protein Y on the gene expression of Enzyme A. Include at which point in gene expression the regulation likely occurs.

      • 1 point for explaining that expression of Enzyme A appears to be stimulated by Protein X and is unaffected by Protein Y. The mRNA, protein, and enzyme activity increased with Protein X, so the regulation must be pretranscriptional.

   ii. **Describe** the influence of Protein X and Protein Y on the gene expression of Enzyme B. Include at which point in gene expression the regulation likely occurs.

      • 1 point for explaining that the expression of Enzyme B appears to be inhibited by Protein X (mRNA, protein, and enzyme activity reduced), so this seems like pre-transcriptional inhibition. The expression of Enzyme B appears to be stimulated by Protein Y. The mRNA and protein levels are the same, but the enzyme level is increased. This implies post-translational activation.

304 | For more free content, visit PrincetonReview.com

AP Biology Premium Prep

D. Another scientist repeats the experiment with Protein Y and Enzyme C and adds a known inhibitor to Protein Y to the tube.

   i. **Predict** the effect on mRNA and protein levels as a result.

- 1 point for stating that the mRNA and protein levels would be unchanged.

   ii. **Predict** the effect on total enzyme activity levels as a result.

- 1 point for stating that the total enzyme activity levels would likely increase to near 18,000.

13. Sample Free-Response Questions | **305**

SAMPLE ESSAY

The following response would earn full points.

A.
  i. Homeostasis is important in all living organisms. In unicellular organisms, temperature must be maintained, osmotic pressure must be controlled, and pH must be regulated at a cellular level.

  - 1 point for describing that unicellular organisms use homeostasis to **regulate temperature, pressure, and pH**

  ii. The same processes are important in multicellular organisms. In addition, a multicellular organism like a plant or an animal has to coordinate its cells to work together. Individual cells must have homeostasis, but the overall living organism must have it too. With more complex systems comes more organisms to regulate like blood pressure and heart rate.

  - 1 point for describing that multicellular organism also use homeostasis to **allow all the cells to work together to support one living organism**

B.

- 1 point for accurate graphing of the data
- 1 point for appropriate axes ranging from approximately **0 to 60,000** on the *y*-axis AND the **three conditions** on the *x*-axis
- 1 point for an appropriate title about **Total Enzyme Levels**, appropriate labels of **enzyme activity levels** on the *y*-axis AND **presence of protein** on the *x*-axis, AND a **key** for the **three genes**
- 1 point for accurate graphing of error bars

C.

i. Gene A expression has no change with the addition of Protein Y, but it has increased transcription, protein, and enzyme activity with the addition of Protein X. It is likely that Protein X is an enhancer of transcription of gene A. Perhaps it binds directly to the promoter region of gene A and assists RNA polymerase to bind. It might also remove a repressor of transcription or work together with other transcription factors. The increase in mRNA would lead to increased protein and increased enzyme activity down the road.

- 1 point for describing that Enzyme A is **stimulated by Protein X** with **pretranscriptional regulation** and **unaffected by Protein Y**

ii. Gene B has a decrease in mRNA, protein, and enzyme activity with Protein X. It has increased total activity with Protein Y. It is likely that Protein X is a repressor of transcription for gene B. It might bind to the promoter of gene B or to another transcription factor to prevent RNA polymerase binding or prevent transcription from occurring in another way. Since the enzyme activity increases with Protein Y, but the mRNA and the protein levels did not change, this means that the enzyme activity was somehow enhanced on the same amount of enzyme that was present without Protein Y. Protein Y might be an enzyme activator of Enzyme B. It might bind to Enzyme B directly or maybe the protein phosphorylates it or cleaves it to make the enzyme active.

- 1 point for describing that Enzyme B is **inhibited by Protein X** with **pretranscriptional regulation** and **stimulated by Protein Z** with **post-translational regulation**

D.

i. The mRNA and the protein would not change with the inhibitor of Protein Y.

- 1 point for predicting that mRNA and protein levels would **not change**

ii. It looks like Enzyme C is unaffected by Protein Y at the mRNA and protein levels, but it looks like the enzyme activity is turned off. With an inhibitor of Protein Y, the enzyme level would likely return to the untreated level of around 18,000.

- 1 point for predicting that the total enzyme activity levels would increase to **approximately 18,000**

HOW TO USE THE SAMPLE FREE-RESPONSE QUESTIONS

The most important advice for this section of the test is *practice, practice, practice!* No matter how well you think you know the material, it's important to practice formulating your thoughts on paper.

You can also access free-response questions from recent exams on the College Board website at apstudents.collegeboard.org/courses/ap-biology/assessment.

Chapter 13 Summary

- o The free-response section features six questions, two of which require long responses. You will have 90 minutes to complete the entire section.

- o Each question is worth a certain number of points. The long-response questions are worth 8–10 points, and the short-response questions are worth 4 points each.

- o Your responses must be written in paragraph form; outlines are not acceptable. Diagrams can be used to supplement your writing but will not receive credit on their own (unless otherwise specified).

- o Remember to keep your response focused, and use specific data and examples (where applicable) to justify your answers.

- o Familiarize yourself with the Key Terms in this book as well as the Big Ideas for the AP Biology course. Understanding these will help you write clear, focused, and accurate responses.

Chapter 14
Laboratory

All AP Biology courses have a laboratory component that gives students hands-on experience regarding some of the biology topics covered in class. Through these laboratory exercises, you can learn the scientific method, lab techniques, and problem-solving skills. There is not an official set of labs that must be completed, and AP teachers get to choose the ones for their classroom. This means that you will not be tested on memorizing specific protocols or data from the labs. However, all labs teach the same basic skills of critical thinking, developing hypotheses, judging results, making conclusions, adapting experiments, and understanding variables. So, those are the skills that you are expected to fully understand. The exam will likely present you with example experiments, and you will need to make decisions and answer questions about them. There are 13 labs that are very popular to do in AP Biology classes, and these are the same labs that often come up on the AP Exam. We have given you short summaries of those 13 labs. This way, if you encounter a question about one of these labs, you will be one step ahead of students who have never even heard of it before.

LAB 1: ARTIFICIAL SELECTION

This lab explores artificial selection—the process by which humans decide traits to enhance or diminish in other species by crossing individuals with the desired phenotype. You need to understand the basic principles of natural selection and how natural selection drives evolution, listed below:

- Variation is present in any population.

- Natural selection, or differential reproduction in a population, is a major mechanism in evolution.

- Natural selection acts on phenotypic variations in populations. Some organisms will have traits that are more favorable to the environment and will survive to reproduce more than other individuals, causing the genetic makeup of the population to change over time.

This lab specifically deals with Wisconsin Fast Plants. In order to artificially select for certain traits, plants with the desired traits can be crossed, changing the genetic makeup of the population. For example, if you wanted to select for height, you could cross only the tallest plants with one another. The new population will have a higher mean height than the previous generation.

LAB 2: MATHEMATICAL MODELING: HARDY-WEINBERG

In order to understand everything you need for the AP Biology Exam, you need to be able to use the Hardy-Weinberg principles and equations to determine allele frequencies in a population. One way to study evolution is to study how the frequencies of alleles change from generation to generation:

- Know how to calculate the allele and genotype frequencies using the two Hardy-Weinberg equations: $p + q = 1$ and $p^2 + 2pq + q^2 = 1$. Don't forget: if the population obeys Hardy-Weinberg's rules, these frequencies remain constant over time.

- Review the discussion of the Hardy-Weinberg principle in this book. Know the five conditions of the Hardy-Weinberg equilibrium: (1) large population, (2) no mutations, (3) no immigration or emigration, (4) random mating, and (5) no natural selection.

- Review natural selection and how it can lead to changes in the genetic makeup of a population.

LAB 3: COMPARING DNA SEQUENCES TO UNDERSTAND EVOLUTIONARY RELATIONSHIPS WITH BLAST

This laboratory uses BLAST (Basic Local Alignment Search Tool), a database that allows you to input a DNA sequence for a gene to look for similar or identical sequences present in other species. In order to understand everything you need for the AP Exam, you need to be able to do the following:

- Look at a phylogenic tree, which is a way to visually represent evolutionary relatedness. Endpoints of each branch correspond to a specific species, and each junction on the tree represents a common ancestor. Species that are closer on a phylogenic tree are more closely related.

- Understand BLAST scoring. Most phylogenic trees are constructed by examining nucleotide sequences; the more identical two species' sequences are for a specific gene, the more closely related they are. BLAST is able to analyze different sequences to tell you how similar they are to one another based on a score. The higher the score, the closer the two sequences align.

LAB 4: DIFFUSION AND OSMOSIS

This lab investigates the process of diffusion and osmosis in a semipermeable membrane as well as the effects of solute concentration and water potential on these processes.

What are the general concepts you really need to know?

Fortunately, this lab covers the same concepts about diffusion and osmosis that are discussed in this book. Just remember that osmosis is the movement of water across a semipermeable membrane, from a region of high water concentration to one of low water concentration, or from a hypotonic region (low solute concentration) to a hypertonic region (high solute concentration).

- Be familiar with the concept of water potential, which is simply the free energy of water. It is a measure of the tendency of water to diffuse across a membrane. Water moves across a selectively permeable membrane from an area of higher water potential to an area of lower water potential.

- Be familiar with the effects of water gain in animal and plant cells. In animals, the direction of osmosis depends on the concentration of solutes both inside and outside the cell. In plants, osmosis is also influenced by turgor pressure—the pressure that develops as water presses against a cell wall. If a plant cell loses water, the cell will shrink away from the cell wall and plasmolyze.

Another important concept to understand is the importance of surface area and volume in cells. There are several questions about each of these topics, but the necessary formulas are listed on the AP Biology Equations and Formulas sheet. Cells maintain homeostasis by regulating the movement of solutes across the cell membrane. Small cells have a large surface area-to-volume ratio; however, as cells become larger, this ratio becomes smaller, giving the cell relatively less surface area to exchange solutes. A cell is limited in size by the surface area-to-volume ratio. There are many organisms that have evolved strategies for increasing surface area, like root hairs on plants and villi in the small intestines of animals.

LAB 5: PHOTOSYNTHESIS

The chemical equation for photosynthesis is

$$6CO_2 + 6H_2O \rightarrow C_6H_{12}O_6 + 6O_2$$

Because plants consume some of this energy during photosynthesis, measuring the oxygen produced by a plant can tell us about the net photosynthesis that is occurring. In this laboratory, photosynthesis rates are measured by using leaf discs that begin to float as photosynthesis is carried out, allowing you to see that photosynthesis is occurring.

There are several properties that affect the rates of photosynthesis, including:

- light intensity, color, and direction
- temperature
- leaf color, size, and type

Be able to hypothesize about the effects of these variables: for example, as light intensity increases, so does the rate of photosynthesis. Remember, both plants and animals contain mitochondria and carry out cell respiration!

LAB 6: CELLULAR RESPIRATION

In this lab, the respiratory rate of germinating and nongerminating seeds and small insects is investigated. The equation for cellular respiration is

$$C_6H_{12}O_6 + O_2 \rightarrow 6CO_2 + 6H_2O$$

Germinating seeds respire and need to consume oxygen in order to continue to grow. Non-germinating seeds do not respire actively. In this lab, the amount of oxygen consumed by these types of seeds is measured with a respirometer. The experiment is also conducted at two temperatures, 25°C and 10°C, because seeds consume more oxygen at higher temperatures. You should know the following:

- Oxygen is consumed in cellular respiration.

- Germinating seeds have a higher respiratory rate than non-germinating seeds, which have a very low respiratory rate.

- Know how to design a study to determine the effect of temperature on cell respiration.

- Know the significance of a control using glass beads. A control is a condition held constant during an experiment. In this case, glass beads are used as a control because they will not consume any oxygen.

LAB 7: MITOSIS AND MEIOSIS

This lab highlights the differences between mitosis and meiosis. In this lab, slides of onion root tips are prepared to study plant mitosis. The important information and skills to review in this lab include the following:

- Mitosis produces two genetically identical cells, while meiosis produces haploid gametes.

- Cell division is highly regulated by checkpoints that depend, in part, on complexes of proteins called cyclins with other proteins called cyclin-dependent kinases. One such example is the mitosis promoting factor (or MPF) that has its highest concentration during cell division and is thought to usher a cell into mitosis.

- Nondisjunction, or the failure of chromosomes to separate correctly, can lead to an incorrect number of chromosomes (too many or too few) in daughter cells.

- Know what each phase of the cell cycle looks like under a microscope. Chapter 7 contains diagrams of each phase.

In one section of this lab, the sexual life cycle of the fungus *Sordaria fimicola* is examined. Sexual reproduction in this fungus involves the fusion of two nuclei—a (+) strain and a (–) strain—to form a diploid zygote. This zygote immediately undergoes meiosis to produce asci, which contain eight haploid spores each.

- During meiosis, crossing-over can occur to increase genetic variation. If crossing-over, or recombination, has occurred, different genetic combinations will be observed in the offspring when compared with the parent strain.

- These offspring with new genetic combinations are called recombinants. By examining the numbers of recombinants with the total number of offspring, an estimate of the linkage map distance between two genes can be calculated with the following equation.

Map distance (in map units) = [(# recombinants) / (# total offspring)] × 100.

LAB 8: BIOTECHNOLOGY: BACTERIAL TRANSFORMATION

In this lab, the principles of genetic engineering are studied. Biotechnologists are able to insert genes into an organism's DNA in order to introduce new traits or phenotypes, like inserting genes into a corn genome that help the crops ward off pests. This process is very complicated in higher plants and animals, but relatively simple in bacteria. You are responsible for knowing the ways in which bacteria can accept fragments of foreign DNA:

- Conjugation: the transfer of genetic material between bacteria via a pilus, a bridge between the two cells

- Transformation: a process in which bacteria take up foreign genetic material from the environment

- Transduction: a process in which a bacteriophage (a virus that infects bacteria) transfers genetic material from one bacteria to another

In addition, DNA can also be inserted into bacteria by using plasmids, which are small, circular DNA fragments that can serve as a vector to incorporate genes into the host's chromosome. Plasmids are key elements in genetic engineering. The concepts you need to know about plasmids include the following:

- One way to incorporate specific genes into a plasmid is to use restriction enzymes, which cut foreign DNA at specific sites, producing DNA fragments. A specific fragment can be mixed together with a plasmid, and this recombinant plasmid can then be taken up by *E. coli*.

- Plasmids can give a transformed cell selective advantage. For example, if a plasmid carries genes that confer resistance to an antibiotic like ampicillin, it can transfer these genes to the bacteria. These bacteria are then said to be transformed. That means if ampicillin is in the culture, only transformed cells will grow. This is a clever way scientists can find out which bacteria have taken up a plasmid.

- In order to make a bacteria cell take up a plasmid, you must (1) add $CaCl_2$, (2) heat shock the cells, and (3) incubate them in order to allow the plasmid to cross the plasma membrane.

LAB 9: BIOTECHNOLOGY: RESTRICTION ENZYME ANALYSIS OF DNA

This laboratory introduces you to the technique of gel electrophoresis. This technique is used in genetic engineering to separate and identify DNA fragments. You need to know the steps of this lab technique for the AP Exam:

1. DNA is cut with various restriction enzymes.

2. The DNA fragments are loaded into wells on an agarose gel.

3. As electricity runs through the gel, the fragments move according to their molecular weights. DNA is a negatively charged molecule; therefore, it will migrate toward the positive electrode. The longer the DNA fragment, the slower it moves through the gel.

4. The distance that each fragment has traveled is recorded.

Restriction mapping allows scientists to distinguish between the DNA of different individuals. Since a restriction enzyme will cut only a specific DNA sequence, it will cause each individual to have a unique set of fragments called restriction fragment length polymorphisms, or RFLPs for short. This technology is used at crime scenes to help match DNA samples to suspects.

LAB 10: ENERGY DYNAMICS

This lab examines energy storage and transfer in ecosystems. Almost all organisms receive energy from the Sun, either directly or indirectly:

- Producers, or autotrophs, are organisms that can make their own food using energy captured from the Sun. They convert this into chemical energy that is stored in high-energy molecules like glucose.

- Consumers, or heterotrophs, must obtain their energy from organic molecules in their environment. They can then take this energy to make the organic molecules they need to survive.

- Biomass is a measure of the mass of living matter in an environment. It can be used to estimate the energy present in an environment.

LAB 11: TRANSPIRATION

This lab investigates the mechanisms of transpiration, the movement of water from a plant to the atmosphere through evaporation. What do you need to take away from this lab?

- The special properties of water that allow it to move through a plant from the roots to the leaves include polarity, hydrogen bonding, cohesion, and adhesion.

- Know the vascular tissues that are involved in transport in plants. Xylem transports water from roots to leaves, while phloem transports sugars made by photosynthesis in the leaves down to the stem and roots.

- Stomata are small pores present in leaves that allow CO_2 to enter for photosynthesis and are also a major place where water can exit a plant during transpiration.

LAB 12: FRUIT FLY BEHAVIOR

In this lab, fruit flies are given the choice between two environments by using a choice chamber, which allows fruit flies to move freely between the two environments. Typically, fruit flies prefer an environment that provides either food or a place to reproduce. They also respond to light and gravity:

- Taxis is the innate movement of an organism based on some sort of stimulus. Movement toward a stimulus is called positive taxis, while movement away from a stimulus is negative taxis. In this lab, fruit flies exhibited a negative gravitaxis (or a movement opposite to the force of gravity) and positive phototaxis (a movement toward a light source).

LAB 13: ENZYME ACTIVITY

This lab demonstrates how an enzyme catalyzes a reaction and what can influence rates of catalysis. In this lab, the enzyme peroxidase is used to catalyze the conversion of hydrogen peroxide to water and oxygen.

$$H_2O_2 + peroxidase \rightarrow 2H_2O + O_2 + peroxidase$$

The following are the major concepts you need to understand for the AP Exam:

- Enzymes are proteins that increase the rate of biological reactions. They accomplish this by lowering the activation energy of the reaction.

- Enzymes have active sites, which are pockets that the substrates (reactants) can enter that are specific to one substrate or set of substrates.

- Enzymes have optimal temperature and pH ranges at which they catalyze reactions. Enzyme concentration and substrate concentration can also influence the rates of reaction.

- If you're asked to design an experiment to measure the effect of these four variables on enzyme activity, keep all the conditions constant except for the variable of interest. For example, to measure the effects of pH in an experiment, maintain the temperature, enzyme concentration, and substrate concentration constant as you change pH.

Stay Up to Date!

That's it! You've reviewed all the content in this book. Now, before you move on to any remaining practice tests, be sure to check the College Board's website for any late-breaking information about test dates, exam formats, or the AP Biology Exam: apstudents. collegeboard.org/courses/ ap-biology/assessment.

Chapter 15
Chapter Drill Answers and Explanations

CHAPTER 4 DRILL: CHEMISTRY OF LIFE

1. **C** Water has many unique properties that favor life, including a high specific heat, surface tension and cohesive properties, polarity, and intermolecular forces due to hydrogen bonding. Eliminate (A), (B), and (D) because they all say water has low values for these properties. Choice (C) correctly describes the high heat capacity of water and the effect this has on a habitat, making (C) the correct answer.

2. **C** Hydrogen bonds are defined as intermolecular forces between the hydrogen portion of one molecule and a fluorine, oxygen, or nitrogen portion of a different molecule. The fluorine, oxygen, or nitrogen must have a lone pair of electrons, which attracts the less electronegative hydrogen atom. The other three bonds listed are covalent bonds, so eliminate (A), (B), and (D).

3. **A** Amino acids are always composed of a hydrogen atom and a functional R-group, also known as a side chain. They are also always composed of a carboxyl group, so eliminate (B) and (D). Finally, they are always composed of an amino group, so eliminate (C).

4. **D** Nucleotides are the monomers of all nucleic acids. Glycerol and fatty acids are the monomers of lipids, so eliminate (A). Monosaccharides are the monomers of carbohydrates, so eliminate (B). Amino acids are the monomers of proteins, so eliminate (C).

5. **D** The bond between the second and third phosphate of ATP makes ATP a good source of energy for functions. The outer phosphate is able to be removed and release energy to be used, turning the molecule into adenosine diphosphate, or ADP. The third phosphate can then be reattached to the molecule, allowing it to be used again for immediate functions.

6. **A** Antacids made from $Mg(OH)_2$ are slightly soluble, so the molecules split into magnesium (Mg^{+2}) ions and hydroxide (OH^{-1}) ions once they reach the stomach. The excess hydroxide ions are attracted to the positively charged hydrogen (H^{+1}) ions, and the two bond to form neutral water molecules, which are excreted via urine. The magnesium ions then pair up with the remaining chloride (Cl^{-1}) ions to form magnesium chloride ($MgCl_2$), which is excreted in feces.

7. **B** All polymers, including polysaccharides such as sucrose, are built through dehydration synthesis reactions. These reactions involve the removal of a water molecule to link two monomers together.

8. **B** Beta-pleated sheets and alpha helices are components of a protein's secondary structure. Primary structure consists of the linear sequence of the amino acids in the polypeptide, so eliminate (A). Tertiary structure consists of the interactions between amino acids that are brought closer together through the folds in the secondary structure, so eliminate (C). Quaternary structure consists of multiple polypeptides interacting with each other, so eliminate (D).

9. **C** Fats, which provide warmth, long-term energy storage, and buoyancy, are made up of lipids. Carbohydrates are used for short-term energy storage, so eliminate (A). Nucleic acids are used for genetic information, so eliminate (B). Proteins are used for a wide variety of functions in living organisms, so eliminate (D).

AP Biology Premium Prep

10. **C** Cholesterol helps hold things together at high temperatures, so eliminate (A). It is not soluble in water, so eliminate (B). It is a hydrophobic molecule, as (C) describes. Its hydrophobic properties allow it to insert into the fatty acid tails that make up cell membranes. While some types of cholesterol are harmful, others play crucial roles, so eliminate (D). Cholesterol helps synthesize hormones and vitamin D.

CHAPTER 5 DRILL: CELL STRUCTURE AND FUNCTION

1. **A** This question is comparing prokaryotic and eukaryotic cells. Since no organized nuclei are present, the sample contains prokaryotic cells. Prokaryotic cells do not have membrane-bound organelle structures, so they do not contain mitochondria, lysosomes, glyoxysomes, Golgi bodies, or chloroplasts. Eliminate (B), (C), and (D). Pigments, such as chlorophyll a, are found throughout the membranes of the cell.

2. **C** Plant, animal, and bacterial cells can be visualized using light microscopy, so eliminate (A), (B), and (D). Cell organelle structures, such as the nuclear membrane, are too small to be observed using light microscopy and require electron microscopy, so the correct answer is (C).

3. **B** The cell would shrink in a hypertonic solution, as water would travel out of the cell into the medium, so eliminate (C). The cell would not change in an isotonic solution or one with an identical osmotic concentration, as water would not travel in or out of the cell; eliminate (A) and (D).

4. **D** The graph shows that the eggplant sample gained approximately 10% mass at 0.1 M, so eliminate (A), (B), and (C). This is because the eggplant cells were placed in a solution that was hypotonic relative to the cells' concentrations. When a plant cell takes in extra mass through passive transport, the extra volume is typically stored in the cells' vacuoles.

5. **B** The eggplant cells will not gain or lose mass when they are in a solution with the same concentration as the cells, or an isotonic solution. The graph shows a 0% change in mass at approximately 0.3 M.

6. **D** Prokaryotic and eukaryotic ribosomes are structurally different and are typically not targeted by the same medications, so eliminate (B). There is no information to support that the antibiotic is not able to cross eukaryotic cell membranes, so eliminate (A). Eukaryotic cells do have ribosomes, but they are much larger than those found in prokaryotic cells, so eliminate (C).

7. **C** A higher surface area to volume ratio makes it easier to maintain temperature as heat can enter and leave the cell more rapidly through the larger surface area, so eliminate (B) and (D). The larger surface area for the cylindrical cell will result in a higher ratio, so eliminate (A).

8. **A** Red blood cells are disc-shaped, not rod-shaped, so eliminate (B). Nerve cells are cylindrical, but they do not travel through spinal fluid, so eliminate (C). Muscle cells are cylindrical, not spherical, so eliminate (D).

15. Chapter Drill Answers and Explanations | **323**

AP Biology Premium Prep

9. **B** The rough endoplasmic reticulum has ribosomes on it, and the ribosomes synthesize proteins that are exported from the cell. The liver produces many enzymes, which are exported to other areas of the body. An amoeba consuming small particles would produce some digestive enzymes, but they wouldn't be made on the rough endoplasmic reticulum, so eliminate (D). A bioluminescent bacteria species does produce enzymes, but it doesn't have rough endoplasmic reticulum; eliminate (A). Xylem cells are involved in the transport of nutrients and water, not proteins; eliminate (C).

CHAPTER 6 DRILL: CELLULAR ENERGETICS

1. **A** Bromothymol blue solution turns yellow when exposed to carbon dioxide, as it becomes acidic. The carbon dioxide is used up during the Calvin cycle of photosynthesis, so the bromothymol blue solution returns to its original blue when it becomes more alkaline.

2. **C** In the Calvin cycle of photosynthesis, carbon is fixed and becomes part of the G3P/PGAL molecule. This molecule is ultimately stored as starch, so the starch will contain the radioactive carbon-14 atoms.

3. **D** Carbonated water is water with dissolved CO_2. When added to the plant, the plant is receiving additional CO_2 and H_2O to the leaves, not the roots, so eliminate (C). Since CO_2 is a reactant for photosynthesis, this allows the plant to grow more. Carbonated water is acidic, not alkaline, and the pH would not be a factor in this situation, so eliminate (A). Oxygen is a product of photosynthesis, not a reactant, so eliminate (B).

4. **B** The seedling's mass increased by incorporating carbon from CO_2. Carbon dioxide is an input for the Calvin cycle. The mass of the soil only marginally decreased, so it couldn't have contributed to the tree's large gain in mass; eliminate (A). Water is used in photosynthesis, but it is used to replace electrons in photosystem II; eliminate (C) and (D).

5. **A** While it is true that the Calvin cycle does not require sunlight, it does require the materials produced in the light-dependent reactions, so the process cannot occur without sunlight; eliminate (D). Carbon dioxide, not oxygen, would be taken in through the stomata overnight; eliminate (C). Oxygen is produced during the light-dependent reactions, which will not operate overnight; eliminate (B).

6. **C** Excess energy lost is expelled as heat energy, which is used to maintain homeostasis by keeping body temperatures consistent in warm-blooded organisms.

7. **A** Glycolysis occurs in the cytoplasm of both eukaryotic and prokaryotic cells. The mitochondrial matrix and membranes are the locations of the Krebs cycle and oxidative phosphorylation, so eliminate (B), (C), and (D).

8. **C** Cellular respiration generates ATP through a series of chemical reactions. The rate of a reaction can be impacted by changes in pressure, reactant concentrations, product concentrations, and temperature. Reduced temperature is the most appropriate factor based on the conditions of hibernation, so eliminate (B) and (D). If cellular respiration decreases, then energy decreases, which will cause the heart rate to decrease as well; eliminate (A).

324 | For more free content, visit PrincetonReview.com

AP Biology Premium Prep

9. **B** To calculate the oxygen consumed per average mass, divide the oxygen consumed by average mass. All four cricket species consumed 0.0002 cm^3/mg of oxygen in three hours, so the answer is (B).

10. **A** The rate of chemical reactions can be measured by observing the rate the reactants are used up or the rate the products are produced. The reactants in enzyme-catalyzed reactions are known as substrates. Enzymes can be reused, so they cannot be used to measure the rates of chemical reactions; eliminate (D). Since products are made in a chemical reaction, they should be increasing; eliminate (C).

CHAPTER 7 DRILL: CELL COMMUNICATION AND CELL CYCLE

1. **B** The enzymes still have to be close together for the signal pathway to occur, even if there are multiple steps; eliminate (A). Heat production is not typically a byproduct of signal transduction pathways, so eliminate (C). A signal transduction pathway typically has multiple unique steps instead of one step being repeated multiple times, so eliminate (D).

2. **A** Different cells have different responses to neurotransmitters and hormones based on their chemical makeup. Cells are not typically permeable to neurotransmitters, and they need to bind to receptors on the surface of the cells; eliminate (B). Smooth muscle cells do contain second messengers, so eliminate (C). Adrenaline only binds to G-protein-coupled receptors, so eliminate (D).

3. **D** Cancer cells have less regulated cell division, which allows them to have higher rates of mitosis; eliminate (B). Cancer cells do not necessarily perform protein translation at a higher rate than healthy cells; eliminate (C). Many cancer cells have shorter cell cycles, which allows them to replicate uncontrollably and take over nearby areas of the body, so the correct answer is (D).

4. **D** Chromosomes from paternal and maternal gametes do not pair up in mitosis, so eliminate (B). Mitosis occurs when cells are replicated and result in two identical copies. The chromosomes are replicated prior to cell division, so eliminate (C). The chromosomes line up in the middle of the cell on the equatorial plate after they've been replicated, so the correct answer is (D).

5. **C** Kinetochores help attach the mitotic spindle, not centromeres; eliminate (A). Centromeres hold sister chromatids together, not nucleosomes; eliminate (B). Nucleosomes, not kinetochores, help package DNA into a compact structure; eliminate (D). Chromatids do distribute one copy of each chromosome to daughter cells to ensure each daughter cell is identical to the original cell, so the correct answer is (D).

6. **A** Translocation may not produce any effect, as the same information exists in the genome, so eliminate (B) and (C). As long as the substitution results in the same amino acid, there will be no identifiable difference in the phenotype, so eliminate (D). Deletion of an enhancer could affect the gene expression and then phenotype, so (A) is correct.

15. Chapter Drill Answers and Explanations | **325**

AP Biology Premium Prep

CHAPTER 8 DRILL: HEREDITY

1. **C** The father and mother are both AB blood type. Their child could have a blood type of A, B, or AB, but not O, so eliminate (D). If you draw a Punnett square, there is a 25% chance of having an AA genotype, a 50% chance of having an AB genotype, and a 25% chance of having a BB genotype. Therefore, the greatest chance is an AB genotype. The answer is (C).

2. **C** This question should be tackled one trait at a time. To create offspring with a 50% chance of having round seeds, the parents must be homozygous recessive and heterozygous, or $Rr \times rr$. To create offspring with a 75% chance of having purple flowers, the parents must both be heterozygotes, or $Pp \times Pp$. $RrPp \times RrPp$ results in 75% of the plants having round seeds; eliminate (A). $RrPp \times rrpp$ results in 50% of the plants having purple flowers; eliminate (B). $RRPp \times rrPp$ results in 100% of the plants having round seeds and 100% of the plants having purple flowers; eliminate (D).

3. **A** Male offspring receive one X-chromosome from their mother and one Y-chromosome from their father. If a mother has the recessive disorder, both of her X-chromosomes must carry the recessive trait. Regardless of which one a son inherits, it will contain the recessive trait. Since males only have one X-chromosome, all sons of an affected mother will be affected by the condition.

4. **D** When two probabilities are separate from each other, the probability of both occurring at the same time in a specific sequence can be found by multiplying the individual probabilities together.

5. **C** X and Z have the lowest crossover frequency, which indicates they are closest together on the chromosome; eliminate (A) and (B). Y and W have the second lowest crossover frequency, so they are next closest together. Y and Z have the highest crossover frequency, so they are the furthest apart; eliminate (D).

6. **C** If a homozygous dominant person reproduces with a homozygous recessive person, all of the offspring will be heterozygous and display the dominant trait. If two people with the heterozygous genotype reproduce, then 75% of their offspring will display the dominant trait (1/3 will be homozygous dominant and 2/3 will be heterozygous) and 25% of the offspring will display the recessive trait (all of them will be homozygous recessive).

7. **A** Gametes have half the number of chromosomes as autosomal cells. A house cat and cheetah's gametes will both have 19 chromosomes, so eliminate (B). A human's gametes will have 23 chromosomes, so eliminate (C). There is no evidence that chromosome number dictates what species can mate, so eliminate (D).

8. **C** Marfan syndrome is an autosomal dominant genetic disorder. All shaded squares and circles represent a homozygous recessive genotype. All open squares and circles represent either a homozygous dominant genotype or a heterozygous genotype. If the individuals in P1 have a chance of having a child without Marfan syndrome, the female individual in P1 must be heterozygous. If a heterozygous individual exhibits a trait, the trait must be dominant. The pattern is not autosomal recessive, because it would be impossible for the two parents in P2 that both have Marfan syndrome to have a child without the disease if it was recessive.

326 | For more free content, visit PrincetonReview.com

AP Biology Premium Prep

9. **B** The snake's genes must be able to respond to environmental stimuli. The heat impacted the phenotype of the snake. Species cannot choose to change their phenotypes, so eliminate (D). Heat energy does not change an organism's genes, so eliminate (C). There is no evidence that scale color indicates a more evolved species, so eliminate (A).

10. **C** Nondisjunction during meiosis could cause two X chromosomes to stay together in an egg gamete. When this egg pairs up with a sperm gamete which also contains an X chromosome, the resulting offspring would have a genotype of XXX. Errors after fertilization would not cause the disorder to appear in a child, so eliminate (D).

CHAPTER 9 DRILL: GENE EXPRESSION AND REGULATION

1. **C** DNA polymerase is an enzyme that pairs nucleotides with complementary base pairs, so the correct answer is (C). Helicase is an enzyme that unwinds the DNA molecule, so eliminate (B). Hydrogen bonds between bases are broken during DNA replication, not sugar-phosphate bonds; eliminate (D). Phosphates are added to the five-prime end of a DNA strand, so eliminate (A).

2. **A** The phosphate groups on DNA molecules give DNA a negative charge, so the correct answer is (A). Histone proteins have a positive charge, so eliminate (D). Nitrogenous bases and polar bonds have negligible or nonexistent charges, so eliminate (B) and (C).

3. **D** DNA ligase connects the discontinued Okazaki fragments together into one DNA strand. DNA polymerase and RNA polymerase add bases to strands of DNA and RNA, respectively, so eliminate (A) and (B). Helicase unwinds the DNA molecule's double helix, so eliminate (C).

4. **C** A base insertion would add an extra base, so eliminate (A). A base substitution would swap one base for another, so eliminate (B). A silent mutation is a type of base substitution that doesn't change the amino acid sequence and doesn't cause a different phenotype, so eliminate (D). The removal of the base at position 508 is known as a base deletion.

5. **C** A duplication occurs when part of a chromosome is replicated and reinserted into itself, resulting in duplicates of gene copies, which is the case here; the correct answer is (C). A deletion occurs when part of a chromosome is removed from itself, so eliminate (A). A translocation occurs when part of a chromosome is detached from itself and attached to a different chromosome, so eliminate (B). An inversion occurs when part of a chromosome is detached and reattached to itself in reverse order, so eliminate (D).

6. **D** Ribosomes synthesize proteins through peptide bonds based on mRNA translation, so eliminate (C). If some proteins are still produced after the antibiotic is administered, there must be some mRNA that was already translated before the drug was applied; the correct answer is (D).

7. **B** The genes that make fireflies glow have been inserted into the genome of the tobacco plant. The tobacco plant is able to perform protein synthesis from this gene and produce the protein that allows fireflies to glow. Because the processes are similar, the gene is able to perform similar functions in both species. Fireflies and tobacco plants are not relatively evolutionarily related, so their proteins

15. Chapter Drill Answers and Explanations | **327**

AP Biology Premium Prep

would not share similar amino acid sequences; eliminate (D). There is no evidence that pathogens are involved or that chloroplasts glow specifically, so eliminate (A) and (C).

8. **A** RNA from the retrovirus is converted into DNA via reverse transcriptase. RNA from the retrovirus can be directly incorporated into the genome, but DNA polymerase and DNA ligase are not the enzymes responsible for this process; eliminate (B) and (D). Enzymes are responsible for this process, so eliminate (C).

9. **D** Individual cells produce unique proteins because specific genes are activated in certain cells based on their location and function. It would be wasteful for all cells to have all genes activated at all times. All autosomal cells contain the same number of chromosomes, so eliminate (A). All cells synthesize some type of protein, so eliminate (B). If genes were deleted from cells' genomes, the cells would likely have significant issues in their function, so eliminate (C).

10. **C** While the nucleotide sequence of DNA provides some input into the role of a cell, its specific function is based on which of those nucleotide sequences, or genes, are activated and expressed in that cell.

11. **A** The polymerase chain reaction generates millions of copies of DNA rapidly, which can be used in other experiments. The DNA is incubated with specific primers, DNA nucleotides, and DNA polymerase enzymes. The process is not perfect, so eliminate (B), and it cannot separate specific segments from other segments or duplicate an entire genome, so eliminate (C) and (D).

CHAPTER 10 DRILL: NATURAL SELECTION

1. **A** Mimicry involves imitating another organism to take advantage of its traits. Coevolution occurs when two species influence each other's evolution, so eliminate (B). Divergent evolution occurs when two species share a common ancestor and deviate into species with similar traits, so eliminate (C). Convergent evolution occurs when two species do not share a common ancestor but develop similar traits due to similar evolutionary pressures, so eliminate (D).

2. **C** Developmental isolation occurs when a zygote does not progress through traditional development and is spontaneously aborted. Temporal isolation occurs when individuals mate at different times of the year, so a zygote is never created; eliminate (A). Geographic isolation occurs when the individuals live in different areas, so a zygote is never created; eliminate (D). Mechanical isolation occurs when the individuals' genitalia are not compatible, so a zygote is never created; eliminate (B).

3. **D** These species are not closely related, but they experienced comparable selection pressures and likely underwent convergent evolution. There is no information to indicate that the predator-prey interactions or mating seasons are similar, so eliminate (B) and (C). They are not closely related, so they don't have similar gene pools; eliminate (A).

4. **B** Sympatric speciation occurs when a new species forms without geographic isolation triggering it. Since there is no geographic isolation mentioned in the question, this is likely the correct process. Allopatric speciation requires geographic isolation, so eliminate (D). X-linked inheritance is irrelevant; this occurs when a trait is found on genes on X-chromosomes. Eliminate (A). Genetic drift occurs when random chance causes evolution in a small population, so eliminate (C).

328 | For more free content, visit PrincetonReview.com

AP Biology Premium Prep

5. **D** Sexual selection occurs when individuals with certain phenotypes are more likely than individuals with other phenotypes to mate and produce offspring. Over time, this can cause distinct differences between genders in a population. Large males tend to be more successful in competing for mates.

6. **D** Choice (D) is the answer because it is the only choice that predicts evolution. Choices (A), (B), and (C) all do not predict evolution. The population is large, natural selection pressures usually exist, and humans usually do not mate randomly. This population will definitely evolve or undergo genetic drift.

7. **B** The origination of the population with one couple is an example of the founder effect. Genetic drift occurs when there is a change in allele frequencies due to random chance. Since this population was small and remote, there was limited gene flow into and out of the population. Natural selection is not a factor as there is no advantage given here; eliminate (C). Sexual selection is not a factor as there is no evidence of differences between genders; eliminate (A) and (D).

8. **A** When antibiotics are first used, there are few bacterial cells with natural resistance. Those cells are not killed by the antibiotics and survive and multiply to create more cells that are resistant. There is no evidence that bacteria transferred from other organisms, mutated to use the antibiotics, or evolved strategies to avoid the antibiotics, so eliminate (B), (C), and (D).

9. **C** Hardy-Weinberg equilibrium is violated when selection or migration occurs. Since the farmer chose to remove a certain phenotype, the population is no longer in Hardy-Weinberg equilibrium.

CHAPTER 11 DRILL: ECOLOGY

1. **C** Since many turtles die quickly after hatching, the curve decreases sharply at the beginning; eliminate (A) and (B). Then, the turtles that reach the ocean have a higher rate of survival, so the curve starts to level off; eliminate (D).

2. **D** The total amount of carbon reacted through photosynthesis, when compared to the amount utilized during cellular respiration, is the net primary productivity. If the rate of cellular respiration is higher than the rate of photosynthesis, then the total biomass of the community is decreasing; eliminate (C). The laws of thermodynamics always apply to all situations, so eliminate (A), and a climax community is irrelevant, so eliminate (B).

3. **B** Different climates in different locations of the world influence the amount and type of vegetation that grows in that area. Precipitation and temperature are two significant factors that influence a climate, so these both cause the changes the journalist observes.

4. **B** While all four are adaptations, the only one related to obtaining energy and nutrients is the claws and teeth of a polar bear. Hibernation and cold-blooded traits help a brown bear and a reptile retain energy, not obtain it, so eliminate (A) and (C). The white fur color of an Arctic rabbit helps the rabbit avoid becoming energy for a predator; eliminate (D).

15. Chapter Drill Answers and Explanations | **329**

AP Biology Premium Prep

5. **A** Since Maui is larger, it can support larger populations than Oahu and can have more individuals; eliminate (B) and (D). There is more likely to be a wide range of habitats on Maui since it is larger, so eliminate (C). The smaller island, Oahu, is likely to have more competition since it is a smaller habitat, so the correct answer is (A).

6. **C** As you move up a trophic level, only 10% of the available energy is passed on to build additional biomass. The rest of the energy is used for metabolism to keep the organism alive. Therefore, each trophic level will get smaller as you move up the pyramid, and the number of individuals at higher trophic levels tends to be smaller.

7. **A** Both forest communities are made up of four different tree species. Community 1 has approximately equal percentages of each species, but Community 2 has a much higher percentage of one species. Therefore, Community 1 has a greater amount of species diversity.

8. **C** Primary and secondary consumers are typically animals, and animals get their nitrogen by preying on animals or plants.

9. **D** The researcher needs to identify if the birds learn their migratory behaviors from other birds or if the migratory behaviors are embedded in their genome. To identify this information, offspring must be removed from parents to see if the offspring still migrate without having been taught the behavior by other birds.

10. **B** If the beavers feed on lily pads as their primary food source, they will decrease when that food source decreases. There is no information to support that lily pads affect cattails or water quality, so eliminate (A) and (D). If the lily pads decrease, coyotes should also decrease as the beavers decrease from losing their food source; eliminate (C).

CHAPTER 12 DRILL: QUANTITATIVE SKILLS AND BIOSTATISTICS

1. **B** In order to answer this question, you must first calculate how much height each tree gained and then divide by the number of trees (in this case, seven).

Tree	Initial Height (inches)	Final Height (inches)	Height Gained (inches)
1	360	372	12
2	320	344	24
3	400	410	10
4	410	425	15
5	375	402	27
6	331	351	20
7	393	405	12

The total height gained is 120 inches divided by 7 trees, which is 17.143 inches. Rounded to the nearest inch, the average height gained is 17 inches.

330 | For more free content, visit PrincetonReview.com

2. **D** Cell A is 10 μm, which is not smaller than 7.5 and is smaller than 48.75; eliminate (B) and (C). In order to determine the median size for all eight cells, organize the sizes in ascending order: 0.1, 2, 2.5, 5, 10, 15, 50, 100. The middle two numbers are 5 and 19; when averaged, this produces a median of 7.5 inches. (D) is correct.

3. **A** The number of eggs released ordered from smallest to largest are:

   100, 107, 107, 110, 156, 209, 220, 221, 301, 307, 333, 390, 407, 411, 455, 500

   The most frequently recurring number in the set above is 107, so this is the mode; eliminate (B) and (D). The smallest number is 100 and the largest is 500. The range is the difference between the two, or 400, so (A) is correct.

4. **B** This question is testing the product rule. A puppy with the required genotype will have brown fur, a short tail, and erect ears. There is a 50% chance of the puppy having brown fur, a 50% chance of the puppy having a short tail, and a 50% chance of the puppy having erect ears. $1/2 \times 1/2 \times 1/2 = 1/8$, which correlates to a 12.5% chance.

5. **C** This question is testing the sum rule, but you also need to use the product rule. A puppy with the first possible genotype will have black fur, a long tail, and floppy ears. There is a 50% chance of the puppy having black fur, a 50% chance of the puppy having a long tail, and a 50% chance of the puppy having floppy ears. $1/2 \times 1/2 \times 1/2 = 1/8$. A puppy with the second possible genotype will have black fur, a short tail, and erect ears. There is a 50% chance of the puppy having black fur, a 50% chance of the puppy having a short tail, and a 50% chance of the puppy having erect ears. $1/2 \times 1/2 \times 1/2 = 1/8$. To find the odds of having either of these genotypes, add the two chances together: $1/8 + 1/8 = 1/4$, or 25%.

6. **D** The first thing you need to do is sort out the alleles for each gene. Since the F_1 generation had red eyes and long wings, you know these alleles must be dominant to the white eyes and short wing alleles. Let's use:

   R = red eyes
   r = white eyes
   L = long wings
   l = short wings

   The parental cross must have been $RRLL \times rrll$, and all F_1 fruit flies were $RrLl$. The null hypothesis (H_0) is that the two genes are segregating independently. This means the expected ratio of F_2 phenotypes will be 9 RL : 3 rL : 3 Rl : 1 rl. Since 500 F_2 flies were generated, you can expect:

   P(red eyes, long wings) = P(RL phenotype) = $9/16 \times 500 = 281.25$

   P(red eyes, short wings) = P(Rl phenotype) = $3/16 \times 500 = 93.75$

   P(white eyes, long wings) = P(rL phenotype) = $3/16 \times 500 = 93.75$

   P(white eyes, short wings) = P(rl phenotype) = $1/16 \times 500 = 31.25$

AP Biology Premium Prep

Next, generate a chart:

F_2 Phenotype	F_2 Genotype	Observed (o)	Expected (e)	$(o - e)^2/e$
Red eyes, long wings	R-L-	190	281.25	29.61
Red eyes, short wings	R-ll	56	93.75	15.20
White eyes, long wings	rrL-	54	93.75	15.20
White eyes, short wings	$Rrll$	200	31.25	911.25
Total		500	500	971.26

In chi-square tests, a calculated χ^2 value is compared to a critical value (CV) from a chi-square table, like the one on the AP Biology Equations and Formulas sheet. If χ^2 is less than the CV, you accept the null hypothesis (H_0). If χ^2 is greater than the CV, you reject H_0. $\chi^2 = 971.26$ and degrees of freedom = # of possibilities − 1 = 4 − 1 = 3, so eliminate (A) and (C). Using $p = 0.01$, you get a critical value of 11.34. Since χ^2 is larger than the critical value, reject H_0. These two genes are not segregating independently, so eliminate (B); (D) is correct.

7.   **A**   Our H_0 is that the observed and expected data match and that the experimental flies have a 3:1 ratio of red to yellow eyes. Based on this, the only possible answer choices are (B) and (C); the other choices mix up how χ^2 values and critical values are compared, so eliminate (B) and (C).

50 flies were analyzed, and we expect three-fourths of them to have red eyes (37.5 flies) and one-fourth of them to have white eyes (12.5 flies). Next, we compare expected (e) and observed (o) data and calculate an χ^2 value.

Phenotype	Observed (o)	Expected (e)	$(o - e)^2/e$
Red eyes	39	37.5	0.06
White eyes, long wings	11	12.5	0.18
Total	50	50	0.24

Because two possibilities are being compared (red and white eyes), the degrees of freedom in this test = 2 − 1 = 1. Using $p = 0.01$, the critical value is 6.63 (which you don't need to look up because it is listed in all answer choices). Since χ^2 is smaller than the CV, you accept H_0. The observed data matches the expected data, and the correct answer is (A).

AP BIOLOGY
EQUATIONS AND FORMULAS

Statistical Analysis and Probability

Mean

$$\bar{x} = \frac{1}{n}\sum_{i=1}^{n} x_i$$

Standard Deviation

$$S = \sqrt{\frac{\sum (x_i - \bar{x})^2}{n-1}}$$

Standard Error of the Mean

$$SE_{\bar{x}} = \frac{S}{\sqrt{n}}$$

Chi-Square

$$\chi^2 = \sum \frac{(o-e)^2}{e}$$

Chi-Square Table

p-value	Degrees of Freedom							
	1	2	3	4	5	6	7	8
0.05	3.84	5.99	7.81	9.49	11.07	12.59	14.07	15.51
0.01	6.63	9.21	11.34	13.28	15.09	16.81	18.48	20.09

Laws of Probability

If A and B are mutually exclusive, then:
$$P(A \text{ or } B) = P(A) + P(B)$$
If A and B are independent, then:
$$P(A \text{ and } B) = P(A) \times P(B)$$

Hardy-Weinberg Equations

$$p^2 + 2pq + q^2 = 1 \qquad p = \text{frequency of allele 1 in a population}$$

$$p + q = 1 \qquad q = \text{frequency of allele 2 in a population}$$

$\bar{x}$ = sample mean

n = size of the sample

s = sample standard deviation (i.e., the sample-based estimate of the standard deviation of the population)

o = observed results

e = expected results

$\sum$ = sum of all

Degrees of freedom are equal to the number of distinct possible outcomes minus one.

Metric Prefixes

Factor	Prefix	Symbol
10^9	giga	G
10^6	mega	M
10^3	kilo	k
10^{-1}	deci	d
10^{-2}	centi	c
10^{-3}	milli	m
10^{-6}	micro	μ
10^{-9}	nano	n
10^{-12}	pico	p

Mode = value that occurs most frequently in a data set

Median = middle value that separates the greater and lesser halves of a data set

Mean = sum of all data points divided by the number of data points

Range = value obtained by subtracting the smallest observation (sample minimum) from the greatest (sample maximum)

Rate and Growth

Rate

$$\frac{dY}{dt}$$

Population Growth

$$\frac{dN}{dt} = B - D$$

Exponential Growth

$$\frac{dN}{dt} = r_{max}N$$

Logistic Growth

$$\frac{dN}{dt} = r_{max}N\left(\frac{K-N}{K}\right)$$

dY = amount of change

dt = change in time

B = birth rate

D = death rate

N = population size

K = carrying capacity

r_{max} = maximum per capita growth rate of population

Simpson's Diversity Index

$$\text{Diversity Index} = 1 - \Sigma\left(\frac{n}{N}\right)^2$$

n = the total number of organisms of a particular species
N = the total number of organisms of all species

Water Potential (ψ)

$$\psi = \psi_P + \psi_S$$

ψ_P = pressure potential

ψ_S = solute potential

The water potential will be equal to the solute potential of a solution in an open container because the pressure potential of the solution in an open container is zero.

The Solute Potential of a Solution

$$\psi_S = -iCRT$$

i = ionization constant (1.0 for sucrose because sucrose does not ionize in water.)

C = molar concentration

R = pressure constant
  (R = 0.0831 liter bars/mole K)

T = temperature in Kelvin (°C + 273)

$$\textbf{pH} = -\log_{10}[H^+]$$

Surface Area and Volume

Surface Area of a Sphere

$$SA = 4\pi r^2$$

Surface Area of a Rectangular Solid

$$SA = 2lh + 2lw + 2wh$$

Surface Area of a Cylinder

$$SA = 2\pi rh + 2\pi r^2$$

Surface Area of a Cube

$$SA = 6s^2$$

Volume of a Sphere

$$V = \frac{4}{3}\pi r^3$$

Volume of a Rectangular Solid

$$V = lwh$$

Volume of a Cylinder

$$V = \pi r^2 h$$

Volume of a Cube

$$V = s^3$$

r = radius

l = length

h = height

w = width

s = length of one side of a cube

SA = surface area

V = volume

Part VI
Additional Practice Tests

- Practice Test 2
- Practice Test 2: Answers and Explanations
- How to Score Practice Test 2
- Practice Test 3
- Practice Test 3: Answers and Explanations
- How to Score Practice Test 3

Practice Test 2

Section I

The Exam

AP® Biology Exam

SECTION I: Multiple-Choice Questions

At a Glance

Total Time
1 hour and 30 minutes
Number of Questions
60
Percent of Total Grade
50%

DISCLAIMER: The official AP Biology Exam will be administered as a hybrid digital exam. Instructions for the digital exam may differ from this practice test.

Instructions

Section I of this examination contains 60 multiple-choice questions.

Indicate all of your answers to the multiple-choice questions on the answer sheet. Give only one answer to each question.

Use your time effectively, working as quickly as you can without losing accuracy. Do not spend too much time on any one question. Go on to other questions and come back to the ones you have not answered if you have time. It is not expected that everyone will know the answers to all the multiple-choice questions.

Section I

AP BIOLOGY
SECTION I
60 Questions
Time—90 minutes

Directions: Each of the questions or incomplete statements below is followed by four suggested answers or completions. Select the one that is best in each case and then fill in the corresponding oval on the answer sheet.

1 ☐ Mark for Review

In general, carbohydrates differ from lipids in that carbohydrates

(A) store energy

(B) are made up of carbon atoms

(C) are macromolecules

(D) are water-soluble

2 ☐ Mark for Review

While completing an experiment similar to the chemical experiment performed by Miller and Urey, a scientist observed that complex organic molecules were formed when precursors were heated and charged with electricity. Which of the following can the scientist conclude?

(A) The Miller-Urey experiment was inaccurate.

(B) The addition of heat allowed the precursors to form more complex organic molecules.

(C) The addition of electricity allowed the precursors to form more complex organic molecules.

(D) The experiment requires a control group without heat or electricity to make a conclusion.

3 ☐ Mark for Review

A chemical agent is found to denature acetylcholinesterase in the synaptic cleft, a space between nerve cells that neurotransmitters cross to cause signaling. What effect will this agent have on the neurotransmitter acetylcholine?

(A) Acetylcholine will not be released from the presynaptic membrane into the synaptic cleft.

(B) Acetylcholine will not bind to receptor proteins on the nerve cell across the synaptic cleft.

(C) Acetylcholine will not diffuse across the cleft to the second cell.

(D) Acetylcholine will not be degraded in the synaptic cleft.

4 ☐ Mark for Review

The base composition of DNA varies from one species to another. Which of the following ratios would you expect to remain constant in the DNA?

(A) Cytosine : Adenine

(B) Pyrimidine : Purine

(C) Adenine : Guanine

(D) Guanine : Deoxyribose

GO ON TO THE NEXT PAGE.

VI. Practice Test 2 | **339**

Section I

Questions 5 through 7 refer to the following.

Consider the following pathway of reactions catalyzed by enzymes (shown in numbers):

$$A \xrightarrow{1} B \xrightarrow{2} C \xrightarrow{3} D \xrightarrow{4} E \xrightarrow{5} F$$
$$\searrow^{6} X \xrightarrow{7} Y$$

5 🔖 Mark for Review

Which of the following situations represents feedback inhibition?

(A) Protein D activating enzyme 4

(B) Protein B stimulating enzyme 1

(C) Protein 7 inhibiting enzyme C

(D) Protein X inhibiting enzyme 2

6 🔖 Mark for Review

An increase in substance F leads to the inhibition of enzyme 3. Which of the following is a direct or indirect result of the process?

(A) A decrease in substance X

(B) Decreased activity of enzyme 6

(C) Decreased activity of enzyme 4

(D) Increased activity of enzyme 5

7 🔖 Mark for Review

If a competitive inhibitor is bound to enzyme 6, which of the following would be decreased?

(A) Protein A

(B) Protein C

(C) Protein X

(D) Protein F

GO ON TO THE NEXT PAGE.

340 | For more free content, visit PrincetonReview.com

Questions 8 through 10 refer to the following.

The affinity of hemoglobin for oxygen is reduced by many factors, including low pH and high CO_2. The graph below shows the different dissociation curves that maternal (normal) hemoglobin and fetal hemoglobin have.

8 Mark for Review

Based on the graph, it can be concluded that

(A) fetal hemoglobin surrenders O_2 more readily than maternal hemoglobin

(B) the dissociation curve of fetal hemoglobin is to the right of maternal hemoglobin

(C) fetal hemoglobin has a higher affinity for O_2 than does maternal hemoglobin

(D) fetal and maternal hemoglobin differ in structure

9 Mark for Review

Which of the following processes would likely shift the normal dissociation curve to the right?

(A) Photosynthesis

(B) Respiration

(C) Fermentation

(D) Mitosis

10 Mark for Review

Hemoglobin's affinity for O_2

(A) decreases as blood pH decreases

(B) increases as H^+ concentration increases

(C) increases as blood pH decreases

(D) decreases as OH^- concentration increases

GO ON TO THE NEXT PAGE.

Section I

11 ☐ Mark for Review

A cell experiences a nondisjunction event in Meiosis I. Calculate the chances of each resulting gamete exhibiting a normal karyotype.

Ⓐ 0%

Ⓑ 25%

Ⓒ 50%

Ⓓ 100%

12 ☐ Mark for Review

G protein-coupled receptors (GPCR) consist of seven transmembrane receptors consisting of alpha helices. The amino end of the protein is on the outside of the plasma membrane. The seven receptors are coupled either on the outside or inside of the plasma membrane. The helix between the 5th and 6th receptors is usually significantly longer than the other helices. A scientist wants to perform an experiment to inhibit the GPCR's function. Which of the following would the scientist aim to target?

Ⓐ The phospholipid bilayer of the plasma membrane

Ⓑ The external side of the receptor

Ⓒ The internal side of the receptor

Ⓓ The sequence of amino acids that make up the binding site for the G protein

13 ☐ Mark for Review

In minks, the gene for brown fur (*B*) is dominant over the gene for silver fur (*b*). Which set of genotypes represents a cross that could produce offspring with silver fur from parents that both have brown fur?

Ⓐ $BB \times BB$

Ⓑ $BB \times Bb$

Ⓒ $Bb \times Bb$

Ⓓ $Bb \times bb$

14 ☐ Mark for Review

Retroviruses violate the Central Dogma because

Ⓐ they store their genetic material as RNA, which is reverse-transcribed to DNA

Ⓑ they do not contain hereditary material but use the host DNA instead

Ⓒ retroviruses do not use the same codon sequences that living organisms do

Ⓓ retroviruses do not require transcription to make proteins

15 ☐ Mark for Review

Which of the following is an example of dehydration synthesis?

Ⓐ Conversion of fats to fatty acids and glycerol

Ⓑ Conversion of proteins to amino acids

Ⓒ Conversion of starch to simple sugars

Ⓓ Conversion of nucleic acids to a strand of DNA

16 ☐ Mark for Review

Which of the following macromolecules can provide short-term energy, speed up chemical reactions, and store genetic information, in that order?

Ⓐ Carbohydrates, proteins, and nucleic acids

Ⓑ Carbohydrates, proteins, and lipids

Ⓒ Carbohydrates, lipids, and proteins

Ⓓ Lipids, proteins, and nucleic acids

GO ON TO THE NEXT PAGE.

For more free content, visit PrincetonReview.com

Questions 17 through 20 refer to the following.

The loop of Henle is a structure within each of the million nephrons within a kidney. As shown in the figure, the two sides have different permeabilities, and there is differential movement across each membrane. The loop acts as a counter-current multiplier that makes the medulla of the kidney very osmotic. The longer the loop, the higher and more powerful the osmolarity gradient that is created. The gradient is required for the reclamation of water from the urine collecting duct. On the right side of the figure is the urine collecting duct. If the body needs to retain water, anti-diuretic hormone makes this region permeable to water via the introduction of aquaporins, and the osmotic pull of the medulla reclaims the water, out of the collecting duct, which makes the urine more concentrated.

17 Mark for Review

Which of the following statements correctly describes the state of things near the top of the descending limb?

(A) The fluid within the descending limb is hypotonic to fluid in the space surrounding the tubule.

(B) The blood within the vasa recta is hypotonic to the filtrate within the descending limb.

(C) The fluid in the area surrounding the tubule is hypertonic to the blood in the vasa recta.

(D) The water in the area surrounding the tubule has a higher water potential than the water in the descending tubule.

18 Mark for Review

What type of transport is occurring when water flows out of the descending tubule?

(A) Simple diffusion

(B) Facilitated diffusion

(C) Active transport

(D) Secondary active transport

19 Mark for Review

If an inhibitor of anti-diuretic hormone, such as caffeine, was ingested, what would be the result?

(A) Aquaporins would appear in the collecting duct.

(B) Aquaporins would increase in number in the collecting duct.

(C) It would block aquaporins like a competitive inhibitor.

(D) Aquaporins would not appear in the collecting duct.

20 Mark for Review

The ability to reclaim water from the collecting duct is directly related to the osmotic pull of the medulla. Kangaroo rats are known to produce extremely concentrated urine. Compared to a human, the nephrons in kangaroo rats must have

(A) thick walls that are impermeable to water

(B) shorter loops of Henle

(C) longer loops of Henle

(D) shorter collecting ducts

GO ON TO THE NEXT PAGE.

Section I

Questions 21 through 22 refer to the following.

The graph below shows two growth curves for bacterial cultures, A and B.

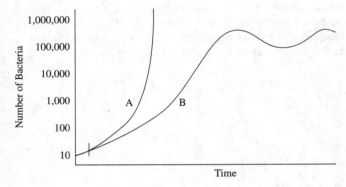

21 Mark for Review

How could a scientist adjust the experiment to have the best chance at increasing the carrying capacity for culture B?

(A) Change the culture's location to a warmer temperature

(B) Decrease the amount of bacteria in the starting culture

(C) Add more growth media to provide additional nutrients

(D) Add samples from culture A to culture B

22 Mark for Review

Which of the following differences was likely used by the scientists when setting up culture A and culture B?

(A) The scientists started culture B with more bacteria than culture A.

(B) The scientists added a competitive inhibitor to culture A.

(C) The scientists did not measure culture B as often.

(D) There was no difference; culture A has not yet exhausted its space and resources.

23 Mark for Review

In plants, the process of phototropism in which an organism responds to a light stimulus would be best exemplified by which of the following?

(A) A sunflower photographed pointing one direction in the morning and another direction at night

(B) A vine climbing and twisting around a trellis

(C) A flowering tree that blooms only in the springtime

(D) A flower that changes color to attract more pollinators

24 Mark for Review

Females with Turner syndrome have a high incidence of hemophilia, a recessive, X-linked trait. Based on this information, it can be inferred that females with this condition

(A) have an extra X chromosome

(B) have an extra Y chromosome

(C) have one less X chromosome than normal

(D) have one less Y chromosome than normal

25 Mark for Review

When a retrovirus inserted its DNA into the middle of a bacterial gene, it altered the normal reading frame by one base pair. This type of mutation is called

(A) duplication

(B) translocation

(C) inversion

(D) frameshift mutation

GO ON TO THE NEXT PAGE.

Section I

26 🔖 Mark for Review

The principal inorganic element or compound found in living things is

(A) carbon

(B) oxygen

(C) water

(D) glucose

27 🔖 Mark for Review

Trisomy X, a disorder in which a female has an extra X chromosome, is the result of nondisjunction. This failure in oogenesis would first be apparent when

(A) the homologous chromosomes are lined up in the middle

(B) the sister chromatids are lined up in the middle

(C) the nuclear envelope breaks down before meiosis

(D) the homologous chromosomes are pulling apart

28 🔖 Mark for Review

Suppose scientists discovered in the fossil record or molecular biology that tiger-like ancestors lived for millions of years before a mutation occurred. This mutation caused a striped phenotype that allowed the animals with this mutation to feed and reproduce at a much faster rate due to this new camouflage adaptation. Which theory of evolution would this discovery support?

(A) Gradualism

(B) Punctuated equilibrium

(C) Genetic drift

(D) Migration

29 🔖 Mark for Review

E. coli have rod-shaped cells that closely resemble cylinders. Which size *E. coli* cell (measured in μm) would be most metabolically efficient?

(A) Radius = 0.5, height = 1.5

(B) Radius = 0.5, height = 2

(C) Radius = 1, height = 3

(D) Radius = 1, height = 4

GO ON TO THE NEXT PAGE.

Section I

Questions 30 through 32 refer to the following.

The following are important pieces of replication and transcription machinery:

30 — Mark for Review

Which of the following figures would be present without helicase?

Ⓐ

Ⓑ

Ⓒ

Ⓓ

31 — Mark for Review

What might have occurred to produce the following situation?

Ⓐ Stalled DNA replication

Ⓑ Initiation of transcription

Ⓒ Repression of transcription

Ⓓ Crossing-over in meiosis

32 — Mark for Review

Put these four situations in the correct order.

Ⓐ I, III, IV, II

Ⓑ I, IV, II, III

Ⓒ IV, I, III, II

Ⓓ I, III, II, IV

GO ON TO THE NEXT PAGE.

33 Mark for Review

Which process is demonstrated in the figure above?

(A) Krebs cycle

(B) Photosynthesis

(C) Glycolysis

(D) Fermentation

34 Mark for Review

Scientists performed a transformation experiment in which they added DNAse, protease, and RNAse to degrade DNA, proteins, and RNA respectively. They found no change in bacteria treated with protease and RNAse, but the transformation did not occur in those treated with DNAse. These results are most consistent with which of the following conclusions?

(A) The double helix is the structure of a DNA molecule.

(B) DNA is the hereditary material.

(C) DNA replication is semiconservative.

(D) DNA polymerase adds bases to only the 3' side of DNA.

35 Mark for Review

Which of the following processes occur in the cytoplasm of a eukaryotic cell?

(A) DNA replication

(B) Transcription

(C) Translation

(D) Translocation

36 Mark for Review

Crossing-over during meiosis permits scientists to determine

(A) the chance for variation in zygotes

(B) the rate of mutations

(C) the distance between genes on a chromosome

(D) which traits are dominant or recessive

37 Mark for Review

An animal cell that is permeable to water but not salts has an internal NaCl concentration of 10%. If placed in freshwater, the cell will

(A) plasmolyze

(B) swell and eventually lyse

(C) endocytose water into a large central vacuole

(D) shrivel

GO ON TO THE NEXT PAGE.

Section I

38 ☐ Mark for Review

Three distinct bird species, flicker, woodpecker, and elf owl, all inhabit a large cactus, *Cereus giganteus*, in the desert of Arizona. Since competition among these birds rarely occurs, the most likely explanation for this phenomenon is that these birds

(A) have a short supply of resources

(B) have different ecological niches

(C) do not live together long

(D) are unable to breed

39 ☐ Mark for Review

Lampreys attach to the skin of lake trout and absorb nutrients from its body. This relationship is an example of

(A) commensalism

(B) parasitism

(C) mutualism

(D) gravitropism

40 ☐ Mark for Review

The nucleotide sequence of a template DNA molecule is 5'-C-A-T-3'. An mRNA molecule with a complementary codon is transcribed from the DNA. What would be the sequence of the anticodon that binds to this mRNA?

(A) 5'-G-T-A-3'

(B) 5'-G-U-A-3'

(C) 5'-C-A-U-3'

(D) 5'-U-A-C-3'

41 ☐ Mark for Review

Viruses are considered an exception to the cell theory because they

(A) require a host cell

(B) do not contain a genome

(C) do not contain a nuclei

(D) cannot evolve via mechanisms that cells evolve from

GO ON TO THE NEXT PAGE.

Section I

Questions 42 through 44 refer to the following.

It is difficult to determine exactly how life began. Answer the following questions as if the following data had been collected billions of years ago.

	4 billion years ago	3.5 billion years ago	3.25 billion years ago	3 billion years ago
Atmospheric carbon dioxide	Present	Present	Present	Present
Atmospheric oxygen	Absent	Absent	Absent	Present
RNA	Absent	Present	Present	Present
Protein	Absent	Absent	Present	Present
DNA	Absent	Absent	Present	Present
Life	Absent	Present	Present	Present

---

42 ▢ Mark for Review

Which best describes the origin of life according to the data?

(A) Life required an environment with atmospheric oxygen and any type of nucleic acids.

(B) Life required an environment with atmospheric carbon dioxide but not atmospheric oxygen.

(C) Life required an environment with self-replicating nucleic acids that can take on many shapes.

(D) Life required an environment with nucleic acids and proteins but not atmospheric oxygen.

---

43 ▢ Mark for Review

How long is the window in which the earliest functional ribosomes could have been found?

(A) 0.10 billion years

(B) 0.25 billion years

(C) 0.50 billion years

(D) 0.75 billion years

---

44 ▢ Mark for Review

Photosynthesis likely began _____ billion years ago when the first _____ appeared.

(A) 4.5; autotrophs

(B) 3.2; autotrophs

(C) 3.5; heterotrophs

(D) 3.2; heterotrophs

GO ON TO THE NEXT PAGE.

VI. Practice Test 2 | **349**

Section I

---

45 Mark for Review

The sequence of amino acids in hemoglobin molecules of humans is more similar to that of chimpanzees than it is to the hemoglobin of dogs. This similarity suggests that

(A) humans and dogs are more closely related than humans and chimpanzees

(B) humans and chimpanzees are more closely related than humans and dogs

(C) humans are related to chimpanzees but not to dogs

(D) humans and chimpanzees are closely analogous

---

46 Mark for Review

Two individuals, one with type B blood, which can be $I^B i$ or $I^B I^B$, and one with type AB blood with genotype $I^A I^B$, have a child. The probability that the child has type O blood, which is caused by genotype ii, is

(A) 0%

(B) 25%

(C) 50%

(D) 100%

Questions 47 through 48 refer to the following.

This bar graph shows the relative biomass of four different populations of a particular food pyramid.

Relative Biomass

Population A
Population B
Population C
Population D

---

47 Mark for Review

How many times greater is the relative biomass of the population with the largest amount of energy available than the population with the smallest amount of energy available?

(A) 1x

(B) 2x

(C) 3x

(D) 5x

---

48 Mark for Review

Which of the following would be the most likely result if there was an increase in the number of organisms in population C?

(A) The biomass of population D would remain the same.

(B) The biomass of population B would decrease.

(C) The biomass of population A would steadily decrease.

(D) The food source available to population C would increase.

GO ON TO THE NEXT PAGE.

Questions 49 through 52 refer to the following.

The cell cycle is a series of events in the life of a dividing eukaryotic cell. It consists of four stages: G_1, S, G_2, and M. The duration of the cell cycle varies from one species to another and from one cell type to another. The G_1 phase varies the most. For example, embryonic cells can pass through the G_1 phase so quickly that it hardly exists, whereas neurons are arrested in the cell cycle and do not divide.

49 Mark for Review

During which phase do chromosomes replicate?

(A) G_1

(B) S

(C) G_2

(D) M

50 Mark for Review

Since neurons are destined never to divide again, what conclusion can be made?

(A) These cells will go through cell division.

(B) These cells will be permanently arrested in the G_0 phase.

(C) These cells will be permanently arrested in the M phase.

(D) These cells will quickly enter the S phase.

51 Mark for Review

If the cell cycle fails to progress, which of the following is a possible explanation?

(A) There are adequate phosphate groups available for the cyclin-dependent kinase.

(B) An oncogene inhibits the signal for apoptosis.

(C) A cyclin is unable to release from its cyclin dependent kinase.

(D) An inhibitor of a cyclin gene has been highly expressed.

52 Mark for Review

In mammalian cells, the first sign of mitosis beginning is the

(A) appearance of chromosomes

(B) separation of chromatids

(C) disappearance of the cell membrane

(D) replication of chromosomes

GO ON TO THE NEXT PAGE.

Section I

Questions 53 through 56 refer to the following.

The figures, which show 5 species of insects that were discovered on a previously unknown island and were named as shown in the table. Proteomic analysis was performed on a highly conserved protein in the insects, and the number of amino acid differences was calculated and included in the table below. The scientists used this data to create the phylogenetic tree shown below with positions labeled I, II, III, and IV as well as O, P, Q, R, and S.

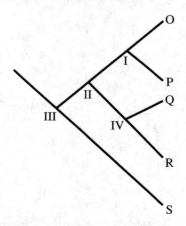

	Species 1	Species 2	Species 3	Species 4	Species 5
Species 1	0	13	3	12	3
Species 2		0	3	8	8
Species 3			0	9	9
Species 4				0	4
Species 5					0

53 Mark for Review

Based on the data, which is the most likely location for Species 1 to be placed?

(A) Position O

(B) Position P

(C) Position R

(D) Position S

54 Mark for Review

Which location contains the oldest common ancestor?

(A) Position I

(B) Position II

(C) Position III

(D) Position IV

55 Mark for Review

Which species is the outgroup?

(A) Species 1

(B) Species 2

(C) Species 3

(D) Species 5

56 Mark for Review

Which 2 species would you expect to have the most shared derived characters?

(A) Species 1 and 5

(B) Species 4 and 5

(C) Species 3 and 4

(D) Species 2 and 3

GO ON TO THE NEXT PAGE.

Section I

Questions 57 through 58 refer to the following.

The data below concerns the general animal body plan of four organisms.

Characteristic	Sea anemone	Hagfish	Eel	Salamander
Vertebral column		+	+	+
Jaws			+	+
Walking legs				+

Note: + indicates a feature that is present in an organism.

57 ☐ Mark for Review

The two most closely related organisms are

(A) the sea anemone and hagfish

(B) the eel and salamander

(C) the hagfish and eel

(D) the sea anemone and salamander

58 ☐ Mark for Review

The correct order of evolution for the traits above is

(A) jaws – vertebral column – walking legs

(B) walking legs – jaws – vertebral column

(C) jaws – walking legs – vertebral column

(D) vertebral column – jaws – walking legs

59 ☐ Mark for Review

Pre- and post-zygotic barriers exist that prevent two different species from producing viable offspring. Which of the following is an example of a post-zygotic barrier to reproduction?

(A) Anatomical differences preventing copulation

(B) Different temporality of mating

(C) Sterility of offspring

(D) Incompatible mating songs

60 ☐ Mark for Review

Birds and insects have both adapted wings to travel by flight. The wings of birds and insects are an example of

(A) divergent evolution

(B) convergent evolution

(C) speciation

(D) genetic drift

STOP

END OF SECTION I

IF YOU FINISH BEFORE TIME IS CALLED, YOU MAY CHECK YOUR WORK ON THIS SECTION. DO NOT GO ON TO SECTION II UNTIL YOU ARE TOLD TO DO SO.

Section II

AP BIOLOGY
SECTION II
6 Questions
Writing Time—90 minutes

DISCLAIMER: The official AP Biology Exam will be administered as a hybrid digital exam. Instructions for the digital exam may differ from this practice test.

Directions: Questions 1 and 2 are long free-response questions that should require about 25 minutes each to answer and are worth 8–10 points each. Questions 3 through 6 are short free-response questions that should require about 10 minutes each to answer and are worth 4 points each.

Read each question carefully and completely. Write your response in the printable free-response booklet found in your online Student Tools. Only material written in the space provided will be scored. Answers must be written out in paragraph form. Outlines, bulleted lists, or diagrams alone are not acceptable unless specifically requested.

---

1 Mark for Review

The human retina has two types of photoreceptors—rods and cones—which respond to light stimuli. Stimulation of photoreceptors is essential for the organisms to respond to visual stimuli. When the rods and cones absorb a photon, a molecule called retinal alters their conformation and allows a sodium channel to close.

The response of rods—that is, the voltage—can be measured with a microelectrode. When a rod cell is bathed in vitreous humor, the fluid found inside the eyeball, it has a resting membrane potential of –40mV. As the rod is exposed to light, the membrane potential becomes more negative, or hyperpolarized.

    A.

        i. **Explain** why the ability to respond to a visual stimulus would be naturally selected for.

        ii. **Identify** whether or not a rod receptor with a resting membrane potential of –65mV would be selected for.

In Experiment 1, a rod was stimulated and its responses measured. Figure 2 shows the hyperpolarizations resulting from two light intensities (two separate stimuli are depicted on the same graph). Figure 3 is a plot of intensity versus hyperpolarization (depicted as a change in mV from –40mV).

In Experiment 2, the Na$^+$ was removed from the vitreous humor. The resting membrane potential changed to –65mV and was unaffected by a light stimulus.

    B. **Explain** how the control in Experiment 1 would differ from the control in Experiment 2.

    C.

        i. **Describe** the relationship between a stimulus and hyperpolarization.

        ii. **Justify** your description.

    D. If the rods were bathed in fluid with a high sodium content, **explain** how the resting membrane potential would change.

GO ON TO THE NEXT PAGE.

Section II

| 2 | Mark for Review |

In a forest there are two types of wildflowers: red flowers and yellow flowers. The red flowers are the food of choice for a population of rodents. The yellow flowers are the food of choice for a type of mountain deer. Both populations of animals appear to be evenly distributed throughout the forest, but it often looks like there are more red flowers on one side of the forest than on the other side.

A. **Explain** how biotic and abiotic factors affect where plants grow.

A researcher wanted to determine if the red flowers preferred to grow on the western side of the forest and the yellow preferred to grow on the eastern side of the forest. For 5 years, she counted the number of red flowers on each side of the forest. The results are shown in Table 1.

Table 1

Year	Western Red Flowers	Eastern Red Flowers
2006	1,527	986
2007	1,324	1,011
2008	1,478	876
2009	1,048	788
2010	1,140	800
Average	1,303	892
Standard Error	208	103

B. **Construct** a graph of averages and standard error for the western and eastern flowers.

C. **Perform** a chi-squared analysis using the average data for both flower types.

D. Using your chi-squared analysis data, **explain** if you can reject the null hypothesis that flowers do not prefer the western side.

GO ON TO THE NEXT PAGE.

Section II

3 Mark for Review

Schizosaccharomyces pombe, or fission yeast, has a cell cycle that resembles that of mammalian cells. During interphase, fission yeast grow to twice their normal size, and at the end of mitosis, both daughter offspring are equal in size to the original parent cell. Genes that regulate the division of fission yeast are known as cell-division cycle, or *cdc*, genes.

The following experiments were conducted to determine the effect of *cdc* gene mutations on yeast cell division.

Experiment 1
In order to determine the effect of *cdc* mutations, wild-type cells and mutants were grown at 37°C in the presence of a radioactive drug that specifically binds to the spindle apparatus. Stages of the cell cycle were elucidated for both cell types incubated at this temperature. At 37°C, temperature-sensitive *cdc* mutants were unable to re-enter interphase after mitosis. The results for wild-type cells and *cdc* mutants are shown in Figure 1.

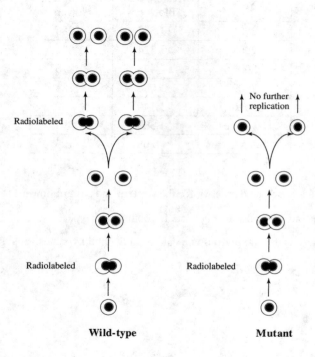

Figure 1

Experiment 2
Cdc mutants were initially incubated at 25°C for 30 minutes, and several rounds of mitosis occurred. Separate colonies were then exposed to temperatures near 37°C at different times during the cell cycle. For each colony, the cell cycle was arrested immediately after the last phase of mitosis.

- A. **Describe** how a mutation in DNA can lead to a protein that no longer functions correctly.

- B. **Explain** the impact of the radiolabeled drug on the cell cycle.

- C. **Predict** what would happen if the *cdc* mutants were initially incubated at 37°C and then switched to 25°C.

- D. **Justify** your prediction.

GO ON TO THE NEXT PAGE.

4. Mark for Review

A prairie was sampled by entomological researchers to determine the species abundance and diversity. Three years into the study a fire swept through the prairie. Two years later quadrant sampling was completed again. Results were compiled in Table 1 below.

Table 1

Organism	Before Fire	2 Years After Fire
Hazel-hued bugs	3	10
Hairy bear moths	14	9
Telluride tiger beetles	6	15
Golden-winged butterflies	21	10

A. **Describe** what happened to the biodiversity of the ecosystem before and after the fire.

B. **Explain** your description using Simpson's Diversity Index.

C. **Predict** what would happen to the biodiversity if there was an invasive insect species introduced into this ecosystem.

D. **Justify** your prediction.

5. Mark for Review

Figure 1 is a pedigree of three generations showing the occurrence of two genetically transmitted diseases. Phenotypes of affected individuals are shown. Circles represent females, and squares represent males. Assume that the diseases are rare, unlinked, and that individuals not blood-related to I-a and I-b ("in-laws") do not have either disease-producing allele.

Figure 1

A. **Explain** how it is possible for a person to have two rare unlinked genetic diseases at the same time.

B. **Calculate** the ratios of affected males and affected females for Disease 2.

C. **Identify** the likely inheritance patterns of Disease 1 and Disease 2.

D. **Describe** how inheritance patterns are shaped by the chromosomes upon which the genes are located.

GO ON TO THE NEXT PAGE.

Section II

---

6 ⬚ Mark for Review

In this fictional scenario, an enzyme was discovered in reptile feces, gertimtonase, that was believed to aid in digestion and water retention. Researchers wanted to know how the enzyme's activity level changed with varying temperature. The scientists took fecal samples of Russian tortoises that were living in varying environmental temperatures. The data was summarized in Figure 1.

Figure 1. Russian tortoise gertimtonase reaction rate with varying temperature

A. **Describe** how you can determine that tortoises raised at 34°C had the highest variability in enzyme reaction rates.

B. Reptiles in the study needed to have a gertimtonase reaction rate exceeding 10 mmol/s for an observed improvement in water retention. **Describe** the range of temperatures that Russian tortoises could be raised in to achieve this benefit.

C. Desert tortoises live in environments where the ground temperature can exceed 60°C. Using the data above, **justify** how the graph would look for desert tortoise gertimtonase.

D. **Explain** at a molecular level how gertimtonase may be affected by varying environmental temperatures.

STOP

END OF EXAM

Practice Test 2: Answers and Explanations

PRACTICE TEST 2 ANSWER KEY

1.	D	21.	C	41.	A
2.	D	22.	D	42.	C
3.	D	23.	A	43.	B
4.	B	24.	C	44.	B
5.	D	25.	D	45.	B
6.	C	26.	C	46.	A
7.	C	27.	D	47.	D
8.	C	28.	B	48.	B
9.	C	29.	A	49.	B
10.	A	30.	C	50.	B
11.	A	31.	C	51.	D
12.	D	32.	D	52.	A
13.	C	33.	B	53.	D
14.	A	34.	B	54.	C
15.	D	35.	C	55.	A
16.	A	36.	C	56.	D
17.	A	37.	B	57.	B
18.	B	38.	B	58.	D
19.	D	39.	B	59.	C
20.	C	40.	C	60.	B

PRACTICE TEST 2 EXPLANATIONS

Section I: Multiple Choice

1. **D** Both carbohydrates and lipids store energy, so eliminate (A). Both lipids and carbohydrates contain carbon and are examples of macromolecules, so eliminate (B) and (C). Carbohydrates are water-soluble and lipids tend to be water-insoluble.

2. **D** The results of the experiment are consistent with the results of the Miller-Urey experiment, so eliminate (A). Choices (B) and (C) are both possible, but with both variables being manipulated at the same time, it is impossible to tell which addition was the crucial one (or both). The scientist needs to perform the experiment without heat or electricity to have a baseline comparison before making a conclusion.

3. **D** This question tests your ability to associate what happens when enzymes are denatured and what would happen in the synaptic cleft. Acetylcholinesterase is an enzyme that degrades acetylcholine in the synaptic cleft. If acetylcholinesterase is denatured, acetylcholine will still be released from the presynaptic membrane (the nerve before the cleft) and continue to diffuse across the synaptic cleft. Acetylcholine will still bind to the postsynaptic membrane (on the nerve after the cleft) because acetylcholine is not degraded. Therefore, you can eliminate (A), (B), and (C).

4. **B** The ratio of purines to pyrimidines should be constant because purines always bind with pyrimidines, no matter which ones they may be.

5. **D** Feedback inhibition occurs when something created by a process then inhibits that process. Choices (A) and (B) involve stimulations, so they can be eliminated. Choice (C) doesn't make sense because it says that "C" is an enzyme, and the passage says that the enzymes are the numbers shown between the steps.

6. **C** If substance F leads to the inhibition of enzyme 3, then substances D and E and enzymes 3, 4, and 5 will be affected. The activity of enzyme 6 will be increased, not decreased, which will therefore increase substance X; eliminate (A) and (B). The activity of enzyme 5 will be decreased, not increased, so eliminate (D).

7. **C** If enzyme 6 were inhibited, then everything that comes after that in the pathway would be reduced. Proteins A and C come before enzyme 6, so eliminate (A) and (B). Protein F will be increased, not decreased, so eliminate (D).

8. **C** Based on the graph, fetal hemoglobin has a higher affinity for oxygen than maternal hemoglobin. Fetal hemoglobin does not give up oxygen more readily than maternal hemoglobin, so eliminate (A). You can also get rid of (B), as the dissociation curve of fetal hemoglobin is to the left of the maternal hemoglobin. Finally, eliminate (D) because while fetal hemoglobin and maternal hemoglobin are different structurally, you can't tell this from the graph.

AP Biology Premium Prep

9. **C** The passage says that high CO_2 and low pH lead to reduced affinity. A right shift curve would represent a reduced affinity. Photosynthesis consumes CO_2, so this would not increase the CO_2; eliminate (A). Mitosis and respiration are not related, so eliminate (B) and (D). Fermentation is done in cells that are doing anaerobic cell respiration, so lactic acid should be released, which would lower pH.

10. **A** Hemoglobin's affinity for O_2 decreases as the concentration of H^+ increases (or the pH decreases) and as the concentration of OH^- increases (or the pH increases). As the pH decreases, the affinity for oxygen will decrease, and as the pH increases, the affinity for oxygen will increase.

11. **A** If meiosis occurred as it should, there should be 4 haploid gametes, each with one long and one short chromosome. When there is a nondisjunction at Meiosis I, the two long homologous chromosomes failed to separate. This causes two gametes to get no copies of the long chromosome and two gametes to get an extra copy. Therefore, there is a 0% chance any of the gametes will have a normal karyotype.

12. **D** The amino acid sequence in the binding site is what alerts the G protein to bind at that location and transmit the message further. If the G protein doesn't bind, the signal molecules will not be conveyed further, and the message will stop.

13. **C** *Bb* and *Bb* is the correct set of genotypes that represents a cross that could produce offspring with silver fur from parents that both have brown fur. Complete a Punnett square for this question. In order for the offspring to have silver fur, both parents must have the silver allele.

14. **A** Retroviruses have an RNA genome (hereditary material, eliminating (B)) that must be reverse-transcribed into DNA (which is the opposite of the Central Dogma first step), and therefore (A) is correct. The codon sequence is used by all things that do translation (like viruses, eliminating (C)). Retroviruses use transcription as well after their genome has been reverse transcribed, so (D) is also incorrect.

15. **D** All of the choices are examples of hydrolysis except the conversion of pyruvic acid to glucose. Hydrolysis is the breaking of a covalent bond by adding water. In (A), (B), and (C), complex compounds are broken down to simpler compounds. The conversion of nucleic acids to DNA is an example of dehydration synthesis. A water molecule is removed as the DNA strand is formed.

16. **A** Carbohydrates provide short-term energy storage, so eliminate (D). Proteins can help speed up chemical reactions as enzymes, so eliminate (C). Nucleic acids store genetic information, so eliminate (B).

17. **A** The figure shows that water leaves the descending limb and flows by osmotic pressure into the surrounding space and then into the vasa recta. This means that each of those places is more hypertonic than the previous. The most hypotonic is the descending tubule, which is why the water leaves.

18. **B** When water flows out of the tubule, it is moving by simple osmosis. However, because it is a polar molecule, it must pass through an aquaporin channel, which is facilitated diffusion.

19. **D** ADH makes the collecting duct permeable to water. If it is inhibited, then it will be unable to do that, meaning (D) is correct. The question doesn't say that the inhibitor is inhibiting the aquaporins directly. Instead, the inhibitor says it is inhibiting the ADH.

20. **C** To make concentrated urine, the medulla must be very osmotic because it must reclaim a lot of water from the collecting duct. To make a very osmotic medulla, the loop of Henle must be very

AP Biology Premium Prep

long. The collecting duct should be very permeable, not impermeable, because concentrated urine means that a lot of water has been reclaimed.

21. **C** The carrying capacity is the maximum number of organisms of a given species that can be maintained in a given environment. Once a population reaches its carrying capacity, the number of organisms will fluctuate around it. By adding more nutrients, the scientist can provide more resources and increase the capacity of the culture. Only (C) discusses adding additional resources to help the culture survive.

22. **D** The growth of curve A is exponential, meaning that the bacteria are replicated as fast as possible without any restrictions. It looks like the populations started at the same level, so eliminate (A). An inhibitor would not cause high growth as shown in culture A, so eliminate (B). Not being measured as often would not cause a graph like the one shown for culture B, so eliminate (C).

23. **A** Phototropism is the ability of a plant to move with regard to the sun. A climbing vine uses pressure to sense the trellis. A tree blooming once a year is using photoperiodism.

24. **C** Females with Turner syndrome have one X chromosome instead of two. If females with this syndrome have a high rate of hemophilia, a likely reason is that they do not have the second X to mask the expression of the disease.

25. **D** This type of mutation is called frameshift mutation. The insertion of DNA leads to a change in the normal reading frame by one base pair. The other answer choices refer to chromosomal aberrations. Choice (A), duplication, occurs when an extra copy of a chromosome segment is introduced. Choice (B), translocation, occurs when a segment of a chromosome moves to another chromosome. Choice (C), inversion, occurs when a segment of a chromosome is inserted in the reverse orientation.

26. **C** The principal inorganic compound found in living things is water. Water is a necessary component for life.

27. **D** An extra X chromosome could be created by a problem separating the homologous chromosomes or sister chromatids. When they are lined up, it would not be apparent that they were not separating properly, but once they start to pull apart, one could see if a mistake had occurred.

28. **B** This scenario describes natural selection, which eliminates (C) and (D). In this situation, the change in tigers is caused by a long period of stasis followed by a major event; this is punctuated equilibrium, (B), and not gradualism, (A).

29. **A** Metabolic efficiency relates directly to the surface area-to-volume ratio. A higher ratio means higher efficiency. Since the ratio is $(2\pi rh + 2\pi r^2) / \pi r^2 h$, this simplifies to $2/r + 2/h$, so if the radius or height is higher, the ratio is smaller. Therefore, the smallest radius and height, (A), gives the highest ratio of 5.33 and the highest efficiency.

30. **C** Helicase is the enzyme that unwinds the helix. The closed helix would exist without helicase.

31. **C** The RNA polymerase is not bound, but there is a bound repressor. The transcription must be repressed.

32. **D** These four phases must be transcription. Transcription begins with a closed helix, so eliminate (C). Then, an enhancer binds, so eliminate (B). Then, RNA polymerase is seen making RNA, so eliminate (A). Finally, the closed helix and the completed mRNA strand exist.

VI. Practice Test 2: Answers and Explanations | **363**

AP Biology Premium Prep

33. **B** Photosynthesis requires light, carbon dioxide, and water and produces oxygen and glucose.

34. **B** The only conclusion possible from this particular experiment is that DNA is the hereditary material. The other conclusions were not tested here.

35. **C** Translation, the synthesis of proteins from mRNA, occurs in the cytoplasm. DNA replication, translocation (the transfer of part of a chromosome to another chromosome), and transcription (the synthesis of RNA from DNA) all occur in the nucleus.

36. **C** Crossing-over permits scientists to determine chromosome mapping. Chromosome mapping is a detailed map of all the genes on a chromosome. The frequency of crossing-over between any two alleles is proportional to the distance between them. The farther apart the two linked alleles are on a chromosome, the more often the chromosome will break between them. Crossing-over does not tell us about the chance of variation in zygotes, the rate of mutations, or whether the traits are dominant, recessive, or masked.

37. **B** In this scenario, the inside of the cell is hypertonic to the outside environment—this will cause water to move into the cell and can cause the cell to lyse.

38. **B** The most likely explanation for this phenomenon is that these birds have different ecological niches. An ecological niche is the position or function of an organism or population in its environment. Eliminate (A) because you do not know if there is a short supply of resources. Choice (C) can also be eliminated because you do not know how long the bird species live together. The breeding patterns of the bird species do not explain the lack of competition, which eliminates (D) as well.

39. **B** This relationship is an example of parasitism. Parasitism is a form of symbiosis in which one organism benefits and the other is harmed. Choice (A), commensalism, is a form of symbiosis in which one organism benefits and the other is unaffected. Choice (C), mutualism, is a form of symbiosis in which both organisms benefit. Choice (D), gravitropism, is the growth of a plant toward or away from gravity.

40. **C** An mRNA molecule uses uracil in place of thymine, so eliminate (A). If the nucleotide sequence of a DNA molecule is 5'-C-A-T-3', then the transcribed DNA strand (mRNA) would be 3'-G-U-A-5'. The nucleotide sequence of the tRNA codon would be 5'-C-A-U-3'.

41. **A** Viruses are considered an exception to the cell theory because they can survive only by invading a host. Choice (B) is incorrect, as all viruses have genomes and some have many genes. Choice (C) is also wrong because prokaryotes are cells, but they do not have nuclei. You can also eliminate (D), as viruses evolve through the same mechanisms that living things do.

42. **C** The origin of life refers to the period when the first life began. Life is shown to be present 3.5 billion years ago when no atmospheric oxygen was present yet, so (A) can be eliminated. Also, there was no life at 4 billion years despite having CO_2 and no O_2; therefore, something else must be required, and (B) can be eliminated. There were no proteins when the first life arose, so (D) is incorrect. Choice (C) states that self-replicating nucleic acids were necessary, which is why life could only occur when RNA appeared.

AP Biology Premium Prep

43. **B** Functional ribosomes would not likely be found before the proteins they make. Therefore, this would be between 3.5 and 3.25 billion years ago. The difference between 3.5 and 3.25 billion years is 0.25 billion years.

44. **B** The sign that photosynthesis began was when oxygen appeared. Oxygen was absent 3.25 billion years ago and present 3 billion years ago, so photosynthesis must have begun between those times; eliminate (A) and (C). Autotrophs, not heterotrophs, perform photosynthesis. Heterotrophs rely on eating other things for their energy.

45. **B** The similarity suggests that humans and chimpanzees are more closely related than humans and dogs. Because these two organisms share similar amino acid sequences, humans must share a more recent common ancestor with chimpanzees than with dogs.

46. **A** There is no way for these two parents to produce a type O child. The only genotype that will result in type O blood is two "O" alleles, one from each parent, since the allele for O blood is recessive to the alleles for A or B blood. The AB parent has alleles for only A and B blood, so it is impossible for these two individuals to produce a child with type O blood.

47. **D** The largest amount of energy is available to the population with the largest biomass, or population B. The smallest amount of energy is available to the population with the smallest biomass, or population D. The bar for population B is approximately 5 times the size of the bar for population D.

48. **B** An increase in the number of organisms in population C would most likely lead to a decrease in the biomass of B because population B is the food source for population C. Make a pyramid based on the biomasses given. If population C increases, population B will decrease. Eliminate (A) and (C), as you cannot necessarily predict what will happen to the biomass of populations that are above population C. Choice (D) can also be eliminated because the food source available to population C would most likely decrease, not increase.

49. **B** Chromosomes replicate during interphase, the S phase. Choices (A) and (C) are incorrect because during G_1 and G_2, the cell makes protein and performs other metabolic duties.

50. **B** Because neurons are not capable of dividing, it is reasonable to conclude that these cells will stay in the G_1 phase (specifically in an extension of G_1 called G_0 phase). This is a reading comprehension question. The passage states that cells that do not divide are arrested at the G_1 phase. Choice (A) is incorrect because these cells will not be permitted to go through cell division. You can also eliminate (C) and (D), as the cells will not enter the M or S phase.

51. **D** To induce cell cycle progression, an inactive CDK (cyclin-dependent kinase) binds a regulatory cyclin. Once together, the complex is activated, can affect many proteins in the cell, and causes the cell cycle to continue. To do this, the CDK transfers phosphate groups onto other molecules. If there are adequate phosphate groups, CDK would function properly and the cell cycle would progress, which rules out (A). Oncogenes keep cells from going through apoptosis, so the cycle would not be stopped, which rules out (B). To inhibit cell cycle progression, CDKs and cyclins are kept separate. Together, they promote the cell cycle. If a cyclin is unable to release from its cyclin-dependent kinase, the cell cycle would be promoted, not stopped, which rules out (C). Choice (D)

VI. Practice Test 2: Answers and Explanations | **365**

AP Biology Premium Prep

is correct because an inhibitor of a cyclin gene would prevent expression of the cyclin gene. With lower levels of the cyclin protein, cell cycle progression would be inhibited.

52. **A** The first sign of prophase in mammalian cells is the appearance of chromosomes, which are usually invisible. They are condensed at the start of mitosis.

53. **D** Since Species 1 has the most differences with the other organisms, it is the out-group, and it must be at position S.

54. **C** Position III, (C), contains the oldest common ancestor because it is the node for all the current species designated by the lettered positions.

55. **A** Species 1, (A), is the out-group because it is the least related to the other species. The table shows that Species 1 has the most differences with all of the other species.

56. **D** Species 2 and Species 3, (D), would have the most shared derived characters because they have the fewest amino acid differences. Species 1 and Species 5 would be expected to have the least since they are so different, which rules out (A). Species 4 and Species 5, (B), were close with only 4 differences, but still not as low as Species 2 and Species 3. Species 3 and Species 4, (C), must be an intermediate, with 9 differences observed in the table.

57. **B** The two most closely related organisms are the two with the most shared derived characteristics.

58. **D** Shared derived characteristics are newly evolved traits that are shared with every group on a phylogenic tree except for one. Vertebral columns are present in every group except for the sea anemone, so it must have evolved first; eliminate (A), (B), and (C). Walking legs are found only in the salamander, indicating that it most likely evolved most recently.

59. **C** Pre-zygotic barriers to reproduction are those that prevent fertilization, so you can eliminate (A), (B), and (D). Choice (C) is an example of a post-zygotic barrier to reproduction.

60. **B** Convergent evolution occurs when two organisms that are not closely related independently evolve similar traits, such as the wings of insects and birds. Divergent evolution occurs when two closely related individuals become different over time and can lead to speciation.

Section II: Free Response

Short student-style responses have been provided for each of the questions. These samples indicate an answer that would get full credit, so if you're checking your own response, make sure that the actual answers to each part of the question are similar to your own. The structure surrounding them is less important, although we've modeled it as a way to help organize your own thoughts and to make sure that you actually respond to the entire question.

Note that the rubrics used for scoring periodically change based on the College Board's analysis of the previous year's test takers. We've done our best to approximate their structure, based on our institutional knowledge of how past exams have been scored and on the information released by the test makers.

Our advice is to over-prepare. Find a comfortable structure that works for you, and really make sure that you're providing all of the details required for each question. Also, continue to check the College Board's website, as they may release additional information as the test approaches. For some extra help, especially if you're worried that you're not being objective in scoring your own work, ask a teacher or classmate to help you out. Good luck!

Question 1

A.

    i. **Explain** why the ability to respond to a visual stimulus would be naturally selected for. (1 point)

If an organism could use visual cues, this could give it an advantage since it might be able to navigate the world more easily. It might access alternative ecological niches or visualize predators or other dangers in the world. Increasing survival will increase evolutionary fitness and having this trait will be selected for.

- 1 point for explaining an **advantage of better visual abilities** in terms of survival

    ii. **Identify** whether or not a rod receptor with a resting membrane potential of –65mV would be selected for. (1 point)

A rod with a low resting membrane potential of –65mV would not be selected for, because when the cell is hyperpolarized to such a large degree, the passage says it does not respond to light stimuli anymore. This means that the visual cues will be absent and the trait won't be selected for.

- 1 point for identifying that a resting membrane potential of –65mV would **NOT be selected for**

B. **Explain** how the control in Experiment 1 would differ from the control in Experiment 2. (2 points)

The control in Experiment 1 would be the rod cell in the vitreous humor that is not stimulated by light. Experiment 2 would have a similar control except that it would be in vitreous humor devoid of sodium and then without stimulation by light.

- 1 point for identifying the control in Experiment 1 as the **rod cell not stimulated by light** and the control in Experiment 2 as the **rod cell in the absence of Na⁺ and not stimulated by light**
- 1 point for explaining the difference between the two controls as the **presence or absence of Na⁺**

C.

    i. **Describe** the relationship between a stimulus and hyperpolarization. (2 points)

When the stimulus was a low light intensity, it caused a small hyperpolarization shown by the small downward bump in Figure 2. However, when the stimulus was a high light intensity, the membrane potential dipped very low, all the way to –65mV.

- 1 point for describing the effect of low light intensity as a **small hyperpolarization** evidenced by the **small dip below -40 mV in Figure 2**
- 1 point for describing the effect of high light intensity as a **large hyperpolarization** evidenced by the **large dip to almost -65 mV in Figure 2**

    ii. **Justify your description. (1 point)**

The intensity of the light stimulus and the level of hyperpolarization have a direct relationship.

- 1 point for identifying the relationships between light intensity and hyperpolarization as **direct**

D. If the rods were bathed in fluid with a high sodium content, **explain** how the resting membrane potential would change. (1 point)

If the sodium content were high in the fluid, it would make the resting membrane potential more positive/less negative, perhaps −20mV. This is the opposite of what happened when the sodium was removed from the vitreous humor. The cell would depolarize at rest because sodium is always flowing through the sodium channel and going into the cell, and the positive sodium ions make it less negative. When a light stimulus occurs, the channel closes, and the cell hyperpolarizes.

- 1 point for explaining how a high sodium content would make the resting membrane potential **more positive** OR **less negative**

Question 2

A. **Explain** how biotic and abiotic factors affect where plants grow. (2 points)

Biotic factors are things that are alive, like plants, fungi, animals, and bacteria. Some of these things compete with the plants, but other things like certain types of fungi and bacteria can help the plant acquire nutrients. Abiotic things are nonliving, like sunlight, soil, nutrients, water, and atmosphere composition. These things are also important, especially since plants need appropriate water and sunlight for photosynthesis.

- 1 point for explaining biotic factors (**living things**) and ONE way they could affect plants (**competition, symbiosis,** etc.)
- 1 point for explaining abiotic factors (**nonliving things**) and ONE way they could affect plants (**resources**)

B. **Construct** a graph of averages and standard errors for the western and eastern flowers. (4 points)

- 1 point for accurate graphing of the data
- 1 point for appropriate axes ranging from approximately **0 to 1600** on the *y*-axis AND the **two flower types** on the *x*-axis
- 1 point for an appropriate title relating **flowers in each location** AND appropriate label of **Average Flowers** on the *y*-axis AND **Location** on the *x*-axis
- 1 point for accurate graphing of error bars

AP Biology Premium Prep

C. **Perform** a chi-squared analysis using the average data for both flower types. (3 points)

Using the null hypothesis that the flowers do not prefer the Western side, we would have expected values of equal numbers living in each place. There were 2,196 flowers total in the average. So, we can expect that there would be 1,098 flowers in each location for the null hypothesis. For the western flowers observed – expected = 1,303 – 1,098 = 205, and for the eastern side observed – expected = 892 – 1,098 = –206. When we square each of them, it is 42,025 and 42,436. Divide each by the expected = 42,025/1,098 and 42, 436/1,098 to get 38.27 for western flowers and 38.64 for eastern flowers. Those get added together to get 76.91 for the chi-squared value.

- 1 point for calculating the expected value of **1,098 flowers** in each location
- 1 point for using the chi-square equation (on the Equations and Formulas sheet) to calculate the values for each location (**38.27** for western flowers and **38.64** for eastern flowers)
- 1 point for adding the two values together to get a chi-square value of **76.91**

D. Using your chi-squared analysis data, **explain** if you can reject the null hypothesis that flowers do not prefer the western side. (1 point)

The chi-square value of 76.91 is larger than the chi-square critical value of 3.84. Therefore, the null hypothesis can be rejected. The flowers appear to grow significantly more on the Western side.

- 1 point for explaining that the **calculated chi-square value is larger than the chi-square critical value** (on the Equations and Formulas sheet), so the null hypothesis should be **rejected**

Question 3

A. **Describe** how a mutation in DNA can lead to a protein that no longer functions correctly. (1 point)

Mutations in DNA are a mistake in the recipe used to make a protein. When the DNA gets expressed, it gets turned into an RNA during transcription. This causes an error in the mRNA. Then, the mRNA gets translated and the codons get read and amino acids get added to build the protein. If there was a mistake in the RNA, then the wrong amino acid would be added or maybe a Stop codon would be added. This can cause the protein to be the wrong shape when it folds up. Protein shape is important for protein function, and if it is the wrong shape, it probably won't work correctly.

- 1 point for describing how **DNA is transcribed into mRNA which is translated into amino acids** which determine a protein and its function; **an error could cause the protein to fold improperly and/or not form at all**

B. **Explain** the impact of the radiolabeled drug on the cell cycle. (1 point)

The radiolabeled drug does not impact the cell cycle. It is only being used to help track the phases of the cell cycle and visualize them. We know the drug has no effect because it is used in the wild-type cells, and it doesn't cause them to have an arrested cell cycle.

- 1 point for explaining that the drug just helps visualize the different stages of the cell cycle and **didn't cause a change in the wild type cells**

VI. Practice Test 2: Answers and Explanations | **369**

AP Biology Premium Prep

C. **Predict** what would happen if the *cdc* mutants were initially incubated at 37°C and then switched to 25°C. (1 point)

If the mutants were started at 37°C, they would not grow. This is what happened in Experiment 1. If they were switched to 25°C, then they would be able to start growing.

- 1 point for predicting that the mutants **would not grow at 37°C AND would grow at 25°C**

D. **Justify** your prediction. (1 point)

I predicted that based on what occurred in Experiment 2. In that experiment they started at the lower temperature, which allowed the cell cycle to proceed. Then at the higher temperature, it caused them to stop the cell cycle. If they begin at the high temp, they won't have a cell cycle, and then when they shift to the lower temperature, it will begin to grow and the cell cycle will continue.

- 1 point for justifying by describing the results of **Experiment 2**

Question 4

A. **Describe** what happened to the biodiversity of the ecosystem before and after the fire. (1 point)

The biodiversity of the ecosystem increased following the fire.

- 1 point for describing that the biodiversity **increased** after the fire

B. **Explain** your description in using Simpson's Diversity Index. (1 point)

Before the fire:

Insect	n	n/N	(n/N)²
Hazel-hued bugs	3	3/44 = 0.068	0.004624
Hairy bear moths	14	14/44 = 0.318	0.101124
Telluride tiger beetles	6	6/44 = 0.136	0.018496
Golden-winged butterflies	21	21/44 = 0.477	0.227529

D = 0.004624 + 0.101124 + 0.018496 + 0.227529 = 0.351773
Simpson's Diversity Index = 1 − D = 0.648227
After the fire:

Insect	n	n/N	(n/N)²
Hazel-hued bugs	10	10/44 = 0.227	0.05129
Hairy bear moths	9	9/44 = 0.205	0.042025
Telluride tiger beetles	15	15/44 = 0.341	0.098596
Golden-winged butterflies	10	10/44 = 0.227	0.05129

AP Biology Premium Prep

$D = 0.05129 + 0.042025 + 0.098596 + 0.05129 = 0.24301$

Simpson's Diversity Index $= 1 - D = 0.756799$

The Simpson's Diversity Index increased, meaning that biodiversity of the ecosystem increased.

- 1 point for explaining by using the Simpson's Diversity Index (on the Equations and Formulas sheet) to calculate the value before the fire as **0.648227**, and the value after the fire as **0.756799**; since **the index increased, so did biodiversity**.

C. **Predict** what would happen to the biodiversity if there was an invasive insect species introduced into this ecosystem. (1 point)

The addition of an invasive species is likely to decrease the biodiversity of the ecosystem.

- 1 point for predicting that the biodiversity will **decrease** after an invasive species is introduced

D. **Justify** your prediction. (1 point)

Since invasive species are unlikely to have a natural predator, they will likely increase in number and consume a large portion of the resources, leading to other organisms being unable to consume those numbers and decreasing in number. Therefore, a greater percentage of the organisms would be made up of the invasive species, while fewer would be made up of the original species, decreasing the biodiversity.

- 1 point for justifying by describing how the **invasive species will likely dominate the environment and resources** because they won't have a predator, causing **native species to die out**

Question 5

A. **Explain** how it is possible for a person to have two rare unlinked genetic diseases at the same time. (1 point)

Everybody has thousands of genes, and it is possible for a person to have mutations in more than one gene. This causes them to have a phenotype that is the combination of all their genes. If a person has the bad luck to have more than one mutated gene, then they could have more than one genetic disease at the same time.

- 1 point for explaining how a person can have **mutations in multiple genes** at the same time

B. **Calculate** the ratios of affected males and affected females for Disease 2. (1 point)

There are 21 males, and 5 are affected = 5/21

There are 20 females, and 0 are affected = 0/20

- 1 point for calculating the male ratio as **5/21** AND the female ratio as **0/20**

C. **Identify** the likely inheritance patterns of Disease 1 and Disease 2. (1 point)

Disease 1 is autosomal dominant, and Disease 2 is X-linked recessive.

- 1 point for identifying Disease 1 as **autosomal dominant** AND Disease 2 as **X-linked recessive**

VI. Practice Test 2: Answers and Explanations | **371**

AP Biology Premium Prep

D. **Describe** how inheritance patterns are shaped by the chromosomes upon which the genes are located. (1 point)

If a gene is located on an X or a Y chromosome, then it will be inherited differently in males and female since males have XY chromosomes, and females have XX chromosomes. This is sex-linked inheritance. In this pedigree with X-linked recessive inheritance, the affected males never have affected fathers. They get the mutated X from their mothers, but the mothers are never affected with the disease themselves. Traits on autosomes get inherited the same in males and females because they each have the same number of autosomes.

- 1 point for describing that **autosomal diseases are inherited equally** by males and females AND that **sex-linked diseases are inherited differently** as **males and females have different sex chromosomes**

Question 6

A. **Describe** how you can determine that tortoises raised at 34°C had the highest variability in enzyme reaction rates. (1 point)

Russian tortoises raised at 34°C had the largest error bars and therefore the largest variability.

- 1 point for describing that tortoises at 34°C had the **largest error bars**

B. Reptiles in the study needed to have a gertimtonase reaction rate exceeding 10 mmol/s for an observed improvement in water retention. **Describe** the range of temperatures that Russian tortoises could be raised at to achieve this benefit. (1 point)

Russian tortoises could be raised between 30°C and 37°C for gertimtonase to have the water retention effect.

- 1 point for describing the range as approximately **30°C to 37°C**

C. Desert tortoises live in environments where the ground temperature can exceed 60°C. Using the data above, **justify** how the graph would look for desert tortoise gertimtonase. (1 point)

The graph's peak would shift to the higher temperatures. Because the tortoise has lived in the desert for many years, its enzymes would be optimized for a higher temperature. Therefore, we would expect to see the peak around 60°C with reaction rates dropping significantly at lower and higher temperatures.

- 1 point for predicting that the peak reaction rate would shift to a **higher temperature** AND for justifying that this would occur since tortoises have **adapted to live at high desert temperatures**

D. **Explain** at a molecular level how gertimtonase may be affected by varying environmental temperatures. (1 point)

Raising temperatures for Russian tortoises to those exceeding 37°C would likely denature the enzyme. Denaturation of the enzyme greatly limits its function, which would lower the reaction rate.

- 1 point for explaining that higher temperatures may **denature** the enzyme, causing it to have little to no functional ability

HOW TO SCORE PRACTICE TEST 2

Section I: Multiple Choice

Section II: Free Response

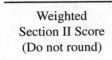

| AP Score Conversion Chart Biology ||
Composite Score Range	AP Score
57–68	5
45–56	4
36–44	3
21–35	2
0–20	1

Composite Score

Practice Test 3

| Section I |

The Exam

AP® Biology Exam

SECTION I: Multiple-Choice Questions

At a Glance

Total Time
1 hour and 30 minutes
Number of Questions
60
Percent of Total Grade
50%

DISCLAIMER: The official AP Biology Exam will be administered as a hybrid digital exam. Instructions for the digital exam may differ from this practice test.

Instructions

Section I of this examination contains 60 multiple-choice questions.

Indicate all of your answers to the multiple-choice questions on the answer sheet. Give only one answer to each question.

Use your time effectively, working as quickly as you can without losing accuracy. Do not spend too much time on any one question. Go on to other questions and come back to the ones you have not answered if you have time. It is not expected that everyone will know the answers to all the multiple-choice questions.

GO ON TO THE NEXT PAGE.

376 | For more free content, visit PrincetonReview.com

Section I

AP BIOLOGY
SECTION I
60 Questions
Time—90 minutes

Directions: Each of the questions or incomplete statements below is followed by four suggested answers or completions. Select the one that is best in each case and then fill in the corresponding oval on the answer sheet.

---

1 ☐ Mark for Review

Acid precipitation is caused by the release of molecules, such as sulfur dioxide and nitrogen oxide, into the atmosphere, where they chemically react with oxygen and water. The products of these reactions are acidic compounds that cause pollution. A scientist wants to propose a strategy to reduce acid precipitation. Which of the following strategies would be most effective?

(A) Restrict the quantities and types of exhaust that factories are able to release into the atmosphere

(B) Require industrial factories to release their exhaust higher in the atmosphere

(C) Use buffers to neutralize large bodies of water with acidic pHs

(D) Implement a purification system that removes heavy metals from industrial exhaust

---

2 ☐ Mark for Review

The Krebs cycle in humans releases

(A) carbon dioxide

(B) pyruvate

(C) glucose

(D) lactic acid

---

3 ☐ Mark for Review

A heterotroph

(A) obtains its energy from sunlight, harnessed by pigments

(B) obtains its energy by catabolizing organic molecules

(C) makes organic molecules from CO_2

(D) obtains its energy by consuming exclusively autotrophs

---

4 ☐ Mark for Review

Regarding meiosis and mitosis, one difference between the two forms of cellular reproduction is that in meiosis

(A) there is one round of cell division, whereas in mitosis there are two rounds of cell division

(B) separation of sister chromatids occurs during the second division, whereas in mitosis separation of sister chromatids occurs during the first division

(C) chromosomes are replicated during interphase, whereas in mitosis chromosomes are replicated during the first phase of mitosis

(D) spindle fibers form during interphase, whereas in mitosis the spindle fibers form during prophase

---

GO ON TO THE NEXT PAGE.

VI. Practice Test 3 | **377**

Section I

5 ⬚ Mark for Review

A feature of amino acids that is NOT found in carbohydrates is the presence of

- Ⓐ carbon atoms
- Ⓑ oxygen atoms
- Ⓒ nitrogen atoms
- Ⓓ hydrogen atoms

6 ⬚ Mark for Review

Which of the following is NOT a characteristic of prokaryotic cells?

- Ⓐ Circular double-stranded DNA
- Ⓑ Membrane-bound cellular organelles
- Ⓒ Plasma membrane consisting of lipids and proteins
- Ⓓ Ribosomes that synthesize polypeptides

7 ⬚ Mark for Review

Which of the following best explains why a population is described as the evolutionary unit?

- Ⓐ Genetic changes can occur only at the population level.
- Ⓑ The gene pool in a population remains fixed over time.
- Ⓒ Natural selection affects individuals, not populations.
- Ⓓ Individuals cannot evolve, but populations can.

8 ⬚ Mark for Review

The endocrine system maintains homeostasis using many feedback mechanisms. Which of the following is an example of positive feedback?

- Ⓐ Infant suckling causes a mother's brain to release oxytocin, which in turn stimulates milk production.
- Ⓑ An enzyme is allosterically inhibited by the product of the reaction it catalyzes.
- Ⓒ When ATP is abundant, the rate of glycolysis decreases.
- Ⓓ When blood sugar levels decrease to normal after a meal, insulin is no longer secreted.

9 ⬚ Mark for Review

A sample of double-stranded DNA from a chimpanzee contains 15,000 base pairs. How many nucleotides are contained in the sample?

- Ⓐ 5,000
- Ⓑ 15,000
- Ⓒ 30,000
- Ⓓ 45,000

GO ON TO THE NEXT PAGE.

For more free content, visit PrincetonReview.com

10 Mark for Review

A large island is devastated by a volcanic eruption. Most of the horses die except for the heaviest males and females of the group. They survive, reproduce, and perpetuate the population. If weight is a highly heritable trait, which statement most likely represents the change to the population before and after the eruption?

Ⓐ A higher mean weight compared to their parents

Ⓑ A lower mean weight compared to their parents

Ⓒ An identical mean weight compared to members of the original population

Ⓓ A higher mean weight compared to members of the original population

11 Mark for Review

Which of the following plays a role in the formation of fingers and toes in human embryos?

Ⓐ Apoptosis

Ⓑ Meiosis

Ⓒ Operons

Ⓓ Plasmids

12 Mark for Review

Life is believed to have begun 3.5 billion years ago. During this time, which of the following gases did the atmosphere on primitive Earth lack?

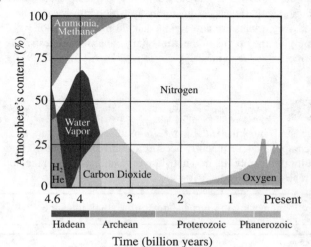

Ⓐ Oxygen

Ⓑ Carbon dioxide

Ⓒ Ammonia

Ⓓ Methane

GO ON TO THE NEXT PAGE.

Section I

Questions 13 through 14 refer to the following.

The digestive system in humans can be divided into two parts: the alimentary canal and the accessory organs. The canal, comprised of the esophagus, stomach, and intestines, is where the food actually passes during its transition into waste. The accessory organs are any organs that aid in the digestion by supplying the organs in the alimentary canal with digestive hormones and enzymes.

13 ☐ Mark for Review

The small intestine is the main site of absorption. It can accomplish it so efficiently because of villi and microvilli that sculpt the membrane into hair-like projections. They likely aid in reabsorption by

(A) increasing the surface area of the small intestine

(B) decreasing the surface area of the small intestine

(C) making the small intestine more hydrophilic

(D) making the small intestine more hydrophobic

14 ☐ Mark for Review

The pancreas is a major accessory organ in the digestive system. Which of the following would destroy the function of the digestive products produced by the pancreas?

(A) A decrease in absorption rates within the alimentary canal

(B) Removing the excess water from the food waste

(C) Increased acidity due to the inability to neutralize stomach acid

(D) An increase in peristalsis and subsequent diarrhea

15 ☐ Mark for Review

In animal cells, which of the following represents the most likely pathway that a secreted protein takes as it is synthesized in a cell?

(A) Plasma membrane–Golgi apparatus–ribosome–secretory vesicle–rough ER

(B) Ribosome–Golgi apparatus–rough ER–secretory vesicle–plasma membrane

(C) Plasma membrane–Golgi apparatus–ribosome– rough ER–secretory vesicle

(D) Ribosome–rough ER–Golgi apparatus–secretory vesicle–plasma membrane

16 ☐ Mark for Review

A scientist analyzes the nucleotides contained in a sample of DNA from a firefly and finds that 30% of the nucleotides are adenine. What percentage of the nucleotides are cytosine?

(A) 20%

(B) 30%

(C) 50%

(D) 70%

17 ☐ Mark for Review

Once specific genes, such as the gene coding for ampicillin, have been incorporated into a plasmid, the plasmid may be used to carry out a transformation, which is

(A) inserting it into a bacteriophage

(B) treating it with a restriction enzyme

(C) inserting it into a suitable bacterium

(D) running a gel electrophoresis

GO ON TO THE NEXT PAGE.

380 | For more free content, visit PrincetonReview.com

Section I

18 ▢ Mark for Review

Although mutations occur at a regular and predictable rate, which of the following statements is a possible reason that the frequency of mutation often *appears* to be low?

- (A) Some mutations produce alleles that are dominant and are always expressed.
- (B) Some undesirable phenotypic traits may be prevented from reproducing.
- (C) Some mutations cause neutral phenotypic changes that linger in the gene pool.
- (D) The predictable rate of mutation results in ongoing variability in a gene pool.

19 ▢ Mark for Review

A scientist wants to test the effect of temperature on seed germination. Which of the following should be part of the experimental design?

- (A) Use temperature as the dependent variable and alter the germination times.
- (B) Use temperature as the independent variable and measure the rate of germination.
- (C) Use temperature as the controlled variable and keep everything identical between groups.
- (D) Use the variable natural outside temperature as a control group.

20 ▢ Mark for Review

A mustard plant seed undergoes a polyploidy event resulting in the new plant's pollen being unable to pollinate the plant that originally produced the seed despite them being less than a meter apart. This best exemplifies which of the following?

- (A) Allopatric speciation because the plants remain in close contact
- (B) Allopatric speciation because there is a significant geographic barrier
- (C) Sympatric speciation because it is the simplest form of speciation
- (D) Sympatric speciation because the plants are not separated

21 ▢ Mark for Review

Which of the following is most correct concerning cell differentiation in vertebrates?

- (A) Cells in different tissues contain different sets of genes leading to structural and functional differences.
- (B) Differences in the timing and expression levels of different genes lead to structural and functional differences.
- (C) Differences in the reading frame of mRNA lead to structural and functional differences.
- (D) Differences between tissues result from spontaneous morphogenesis.

GO ON TO THE NEXT PAGE.

VI. Practice Test 3 | **381**

Section I

Questions 22 through 23 refer to the following.

Pumping blood through the human heart must be carefully organized for maximal efficiency and to prevent backflow. In the figure below, the blood enters the heart through the vena cava (1), passes through the right atrium and right ventricle and then goes through the pulmonary artery toward the lungs. After traveling through the lungs, the blood returns through the pulmonary vein and then passes into the left atrium and the left ventricle before leaving the heart via the aorta.

22 Mark for Review

Which of the following chambers or vessels carry deoxygenated blood in the human heart?

(A) 1 only

(B) 2 and 3

(C) 1, 2, 3, 4

(D) 4 and 5

23 Mark for Review

Blood is pumped via heart contractions triggered by action potentials spreading through the heart muscle. If there is a sudden increase in blood in chamber 3, which chamber of the heart received an increased number of action potentials?

(A) Left atrium

(B) Left ventricle

(C) Right atrium

(D) Right ventricle

24 Mark for Review

Some strains of viruses can change normal mammalian cells into cancer cells in vitro. Which of the following is the best explanation for this impact on the mammalian cell?

(A) A pilus is formed between the mammalian cell and the virus.

(B) The viral genome incorporates into the mammalian cell's nuclear DNA.

(C) The host's genome is converted into the viral DNA.

(D) There is a viral release of spores into the mammalian cell.

25 Mark for Review

Which of the following statements correctly describes meiosis?

(A) Meiosis produces four diploid gametes.

(B) Homologous chromosomes join during meiosis II.

(C) Sister chromatids separate during meiosis I.

(D) Crossing-over increases genetic variation in gametes.

GO ON TO THE NEXT PAGE.

26 Mark for Review

A scientist wants to explore events that can prevent interspecies breeding. Which of the following breeding combinations would allow the scientist to investigate barriers to interspecies breeding?

Ⓐ A male and female from two different species of finches that live on different islands

Ⓑ Two female sea turtles from the same species

Ⓒ Two male blue-footed boobies that perform a mating dance

Ⓓ A male and female bear from the same species with different colored fur

27 Mark for Review

Which of the following is a characteristic of asexual reproduction in animals?

Ⓐ Progeny cells have a different number of chromosomes from the parent cell.

Ⓑ Progeny cells are identical to the parent cell.

Ⓒ The parent cell produces haploid cells.

Ⓓ The progeny cells require another organism to form a zygote.

28 Mark for Review

Transpiration is a result of special properties of water. Which of the following is a special property of water that makes transpiration possible?

Ⓐ Nonpolarity

Ⓑ Adhesion

Ⓒ High heat capacity

Ⓓ Hydrophobicity

GO ON TO THE NEXT PAGE.

Section I

Questions 29 through 32 refer to the following.

An experiment was performed to assess the growth of two species of plants when they were grown in different pHs, given different volumes of water, and watered at different times of day over 6 weeks. Two plants were grown of each species and the average heights (in cm) are shown in the table.

		Species A	Species B
pH	2	3.2	4.1
	4	37.6	20.6
	7	62.3	22.4
	10	48.4	13.5
	13	4.1	2.7
Volume (mL)	10	4.9	12.4
	20	19.2	38.9
	40	56.2	45.6
	80	65.1	21.5
	160	2.6	1.8
Time	12:00 A.M.	62.3	20.3
	7:00 A.M.	61.1	21.8
	12:00 P.M.	66.7	18.4
	7:00 P.M.	65.3	19.3

29 ☐ Mark for Review

For which condition(s) do the species have different preferences?

(A) pH

(B) Volume

(C) Volume and watering time

(D) pH and volume and watering time

30 ☐ Mark for Review

What are the preferred growth conditions for Species B?

(A) pH 7, 40 mL, any time of day

(B) pH 10, 40 mL, 7:00 A.M.

(C) pH 7, 80 mL, any time of day

(D) pH 10, 80 mL, 12:00 P.M.

31 ☐ Mark for Review

Which pH and volume were likely used for the watering time experiment?

(A) pH 4 and 40 mL

(B) pH 7 and 40 mL

(C) pH 4 and 80 mL

(D) pH 7 and 80 mL

GO ON TO THE NEXT PAGE.

384 | For more free content, visit PrincetonReview.com

Section I

32 ☐ Mark for Review

Which of the following would most improve the statistical significance of the results?

Ⓐ Let the plants grow for a longer period of time

Ⓑ Add more conditions to test, such as amount of light and amount of soil

Ⓒ Test the same plants with more pHs, volumes, and times of day

Ⓓ Increase the number of plants in each group

33 ☐ Mark for Review

Photoperiodism in plants, in which plants respond to the stimulus of the day lengthening or shortening, can be best compared to which of the following phenomena in animals?

Ⓐ Viral infection

Ⓑ Increased appetite

Ⓒ Meiotic cell division

Ⓓ Circadian rhythms

34 ☐ Mark for Review

In most ecosystems, gross primary productivity—the total amount of chemical energy that producers create in a given time—is not entirely available for the consumers to utilize due to which explanation below?

Ⓐ Not all solar energy is in the correct spectrum for plants to absorb it.

Ⓑ Plants utilize some energy for the cellular respiration.

Ⓒ Heterotrophs do not absorb energy from autotrophs.

Ⓓ Very little biomass is available at the producer level.

35 ☐ Mark for Review

Hawkmoths are insects that are similar in appearance and behavior to hummingbirds. Which of the following is a valid explanation for these similarities?

Ⓐ These organisms are examples of convergent evolution.

Ⓑ These organisms were subjected to different environmental conditions.

Ⓒ These organisms are genetically related to each other.

Ⓓ These organisms have homologous structures.

36 ☐ Mark for Review

Which of the following describes a mutualistic relationship?

Ⓐ A tapeworm feeds off its host's nutrients causing the host to lose large amounts of weight.

Ⓑ Certain plants grow on trees in order to gain access to sunlight without affecting the tree.

Ⓒ Remora fish eat parasites off sharks. The sharks stay free of parasites, and the remora fish are protected from predators.

Ⓓ Meerkats sound alarm calls to warn other meerkats of predators.

GO ON TO THE NEXT PAGE.

VI. Practice Test 3 | **385**

Section I

37 ☐ Mark for Review

The pancreas is an organ that makes insulin and glucagon in its beta and alpha cells, respectively. Insulin is released when blood glucose is high, and glucagon is released when blood glucose is low. Anti-beta cell antibodies, which bind to their target and inhibit functionality, will cause which of the following to occur?

(A) Glucagon secretion will stop, and blood glucose levels will not decrease.

(B) Glucagon secretion will stop, and blood glucose levels will decrease.

(C) Glucagon secretion will stop, and digestive enzymes will be secreted.

(D) Insulin secretion will stop, and blood glucose levels will not decrease.

Questions 38 through 39 refer to the following.

The table below shows Watson-Crick base pairing (white) and wobble pairing (shaded) for RNA. There is even another nucleotide base that appears in tRNA anticodons. The wobble pairing can be seen between the nucleotide in the third position on the anticodon and the nucleotide in the 3' most position on an mRNA codon.

anticodon base			codon base
	CG	GU	
	AU	UG	
	UA	IU	
	GC	IA	
		IC	

C: cytosine; A: adenine; G: guanine; U: uracil; I: inosine.

38 ☐ Mark for Review

Which of the following is the result of wobble pairing?

(A) There are several possible codons for a given anticodon.

(B) There are more amino acids than there are nucleotide bases.

(C) There are more inosines in mRNA than there are in tRNA.

(D) Some codons will have only one possible anticodon.

39 ☐ Mark for Review

If the mRNA and the tRNA are oriented in an antiparallel direction during translation, what position on the tRNA is the wobble position?

(A) Always on the 5' end of the anticodon

(B) Always on the 3' end of the anticodon

(C) Sometimes on the 5' end and sometimes on the 3' end

(D) Neither; tRNA is not linear like mRNA

GO ON TO THE NEXT PAGE.

386 | For more free content, visit PrincetonReview.com

Section I

Questions 40 through 41 refer to the following.

A scientist wanted to study how the enzyme catalase breaks down hydrogen peroxide. Catalase was extracted from vegetables by pureeing them with either 20°C water or 100°C water. The scientist made 2% and 16% hydrogen peroxide solutions. Filter paper was soaked in the blended catalase and placed into containers with the hydrogen peroxide solutions. The scientist measured the number of seconds, up to 100, it took for the disks to float to the top of the hydrogen peroxide solutions due to the bubbles produced in order to measure the enzyme's activity. The data is in the table below.

Concentration of H_2O_2 Solution	Temperature of Catalase	Average Time to Float (s)
2%	20°C	45
2%	100°C	100+
16%	20°C	20
16%	100°C	100+

40 ▢ Mark for Review

What happened to the catalase enzyme that caused it to stop functioning at the higher temperature?

(A) The catalase enzyme did not stop functioning at any point.

(B) The catalase enzyme worked to its full capacity and hit a limit of how much product it could produce per second.

(C) The reaction ran out of substrate to continue producing bubbles.

(D) The catalase enzyme's structure was changed, so the substrate was no longer able to bind.

41 ▢ Mark for Review

According to the data, does catalase have greater functionality at higher or lower temperatures and concentrations?

(A) Higher concentrations and lower temperatures

(B) Higher concentrations and higher temperatures

(C) Lower concentrations and lower temperatures

(D) Lower concentrations and higher temperatures

42 ▢ Mark for Review

The calypso orchid, *Calypso bulbosa*, grows in close association with mycorrhizae fungi. The fungi penetrate the roots of the flower and take advantage of the plant's food resources. The fungi concentrate rare minerals, such as phosphates, in the roots and make them readily accessible to the orchid. This situation is an example of

(A) parasitism

(B) commensalism

(C) mutualism

(D) endosymbiosis

43 ▢ Mark for Review

The synthesis of new proteins necessary for lactose utilization by the bacterium *E. coli* using the *lac* operon is regulated by the ability of RNA polymerase to bind and advance. This regulation can best be described as

(A) bacterial regulation

(B) pre-transcriptional regulation

(C) pre-translational regulation

(D) post-translational regulation

44 ▢ Mark for Review

Trypsin is a digestive enzyme. It cleaves polypeptides after lysine and arginine amino acid residues. Which of the following statements about trypsin is NOT true?

(A) It is an organic compound made of proteins.

(B) It is a catalyst that alters the rate of a reaction.

(C) It is operative over a wide pH range.

(D) The rate of catalysis is affected by the concentration of substrate.

GO ON TO THE NEXT PAGE.

VI. Practice Test 3 | **387**

Section I

45 ☐ Mark for Review

In DNA replication, which of the following occurs?

(A) Topoisomerase unwinds the double helix.

(B) DNA ligase links the Okazaki fragments.

(C) RNA polymerase is used to elongate both chains of the helix.

(D) DNA strands grow in the 3' to 5' direction.

46 ☐ Mark for Review

Which of the following is true about genetic variation?

(A) Mutation is the greatest source of genetic variation.

(B) Ecosystems with high levels of genetic variation are not resistant to stress.

(C) Mitosis provides genetic variation to most somatic cells.

(D) Crossing-over prevents plants from undergoing speciation.

47 ☐ Mark for Review

Approximately how many ATP will be generated from the electron transport chain from the two pyruvate molecules that enter the Krebs cycle?

(A) 2

(B) 10

(C) 15

(D) 18

48 ☐ Mark for Review

If a plant undergoing the light-dependent reactions of photosynthesis began to release $^{18}O_2$ instead of normal oxygen, one could most reasonably conclude that the plant had been supplied with

(A) H_2O containing radioactive oxygen

(B) CO_2 containing radioactive oxygen

(C) $C_6H_{12}O_6$ containing radioactive oxygen

(D) NO_2 containing radioactive oxygen

49 ☐ Mark for Review

Yeast haploid cells secrete pheromones to other yeast to indicate they want to mate, and others respond by growing toward these potential mates. The pheromones bind to a receptor, which eventually leads to increased expression of transcription factors required for this growth. This best exemplifies which of the following?

(A) Chemical inhibition due to the intracellular binding site

(B) Passive transport since the haploid cells require less energy

(C) Mitotic division due to anaphase elongation

(D) Signal transduction affecting cell function

GO ON TO THE NEXT PAGE.

Section I

50 — Mark for Review

Homologous structures are often cited as evidence for the process of natural selection. Which of the following is an example of a pair of analogous structures?

(A) The forearms of a cat and the wings of a bat

(B) The flippers of a whale and the arms of a man

(C) The pectoral fins of a porpoise and the flippers of a seal

(D) The forelegs of an insect and the forelimbs of a dog

51 — Mark for Review

Certain populations of finches have long been isolated on the Galápagos Islands off the western coast of South America. Compared with the larger stock population of mainland finches, these separate populations exhibit far greater variation over a wider range of species. The variation among these numerous finch species is the result of

(A) convergent evolution

(B) divergent evolution

(C) disruptive selection

(D) stabilizing selection

52 — Mark for Review

Which of the following contributes the MOST to genetic variability in a population?

(A) Sporulation

(B) Binary fission

(C) Vegetative propagation

(D) Mutation

GO ON TO THE NEXT PAGE.

VI. Practice Test 3 | **389**

Section I

Questions 53 through 55 refer to the following.

A marine ecosystem was sampled in order to determine its food chain. The results of the study are shown below.

Type of Organism	Number of Organisms
Shark	2
Small crustaceans	400
Mackerel	20
Phytoplankton	1,000
Herring	100

53 ☐ Mark for Review

Which of the following organisms in this population are secondary consumers?

(A) Sharks

(B) Phytoplankton

(C) Herrings

(D) Small crustaceans

54 ☐ Mark for Review

Which of the following organisms has the largest biomass in this food chain?

(A) Phytoplankton

(B) Mackerels

(C) Herrings

(D) Sharks

55 ☐ Mark for Review

If the herring population is reduced by predation, which of the following would most likely be a secondary effect on the ecosystem?

(A) The mackerels will become the largest predator in the ecosystem.

(B) The small crustacean population will be greatly reduced.

(C) The phytoplankton population will be reduced over the next year.

(D) The small crustaceans will become extinct.

GO ON TO THE NEXT PAGE.

Questions 56 through 57 refer to the following.

Scientists used embryology, morphology, paleontology, and molecular biology to create the phylogenetic tree below.

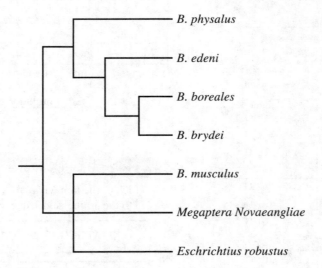

56 — Mark for Review

Which of the following is the LEAST closely related according to the phylogenetic tree?

(A) *B. physalus* and *B. brydei*

(B) *B. boreales* and *B. brydei*

(C) *B. musculus* and *B. brydei*

(D) *B. musculus* and *Eschrichtius robustus*

57 — Mark for Review

Which of the sources of evolutionary evidence would be the most reliable of those listed?

(A) Embryology

(B) Morphology

(C) Paleontology

(D) Molecular biology

58 — Mark for Review

The brain of a frog is destroyed. A piece of acid-soaked paper is applied to the frog's skin. Every time the piece of paper is placed on its skin, one leg moves upward. Which of the following conclusions is best supported by the experiment?

(A) Reflex actions are not automatic.

(B) Some reflex actions can be inhibited.

(C) All behaviors in frogs are primarily reflex responses.

(D) This reflex action does not require the brain.

GO ON TO THE NEXT PAGE.

Section I

Questions 59 through 60 refer to the following.

First Position (5' end)	Second Position	The Genetic Code: Codons of mRNA that Specify a Given Amino Acid			
		Third Position (3' end)			
		U	C	A	G
U	U	UUU UUC Phenylalanine		UUA UUG Lucine	
	C	UCU UCC UCA UCG Serine			
	A	UAU UAC Tytrosine		UAA	UAG
	G	UGU UGC Cysteine		UGA	UGG Tryptophan
C	U	CUU CUC CUA CUG Leucine			
	C	CCU CCC CCA CCG Proline			
	A	CAU CAC Histidine		CAA CAG Glutamine	
	G	CGU CGC CGA CGG Arginine			
A	U	AUU AUC AUA Isoleucine			AUG
	C	ACU ACC ACA ACG Threonine			
	A	AAU AAC Asparagine		AAA AAG Lysine	
	G	AGU AGC Serine		AGA AGG Arginine	
G	U	GUU GUC GUA GUG Valine			
	C	GCU GCC GCA GCG			
	A	GAU GAC Aspartic Acid		GAA GAG Glutamic acid	
	G	GGU GGC GGA GGG Glycine			

mRNA: A U G C C A C U A G C A C G U

| Met | Pro | Leu | Ala | Arg | Protein |
| Methionine | Proline | Leucine | Alanine | Arginine | |

Formation of a Protein

59 ☐ Mark for Review

Which of the following DNA strands is the template strand that led to the amino acid sequence shown above?

(A) 3'-ATGCGACCAGCACGT-5'

(B) 3'-AUGCCACUAGCACGU-5'

(C) 3'-TACGGTGATCGTGCA-5'

(D) 3'-UACGGUGAUCGUGCA-5'

60 ☐ Mark for Review

Immediately after the translation of methionine, a chemical is added which deletes all remaining uracil nucleotides in the mRNA. Which of the following represents the resulting amino acid sequence?

(A) Serine–histidine–serine–threonine

(B) Methionine–proline–glutamine–histidine

(C) Methionine–proline–leucine–alanine–arginine

(D) Methionine–proline–alanine–arginine–arginine

STOP

END OF SECTION I

IF YOU FINISH BEFORE TIME IS CALLED, YOU MAY CHECK YOUR WORK ON THIS SECTION. DO NOT GO ON TO SECTION II UNTIL YOU ARE TOLD TO DO SO.

Section II

AP BIOLOGY
SECTION II
6 Questions
Writing Time—90 minutes

DISCLAIMER: The official AP Biology Exam will be administered as a hybrid digital exam. Instructions for the digital exam may differ from this practice test.

Directions: Questions 1 and 2 are long free-response questions that should require about 25 minutes each to answer and are worth 8–10 points each. Questions 3 through 6 are short free-response questions that should require about 10 minutes each to answer and are worth 4 points each.

Read each question carefully and completely. Write your response in the printable free-response booklet found in your online Student Tools. Only material written in the space provided will be scored. Answers must be written out in paragraph form. Outlines, bulleted lists, or diagrams alone are not acceptable unless specifically requested.

---

1   🔖 Mark for Review

A researcher is studying *Asterina stellaris*, a species of starfish that lives in the tidal pools of North America. She gathered data on the number of starfish found within a twenty-meter square plot over the course of a year. The data is shown in Table 1.

Table 1. Distribution of Starfish within a Twenty-Meter Square Plot

Month of Observation	Average Distance Between Starfish (inches)
January	9.0
March	17.2
May	29.1
July	27.5
September	19.6
November	11.1

A.   **Describe** the pattern of distribution of the starfish using the provided data. Be specific as to how the data changes based on the time of year.

B.   **Identify** three physical or ecological conditions that could cause the starfish to change their distance from other starfish.

C.   **Explain** how each condition could bring about the pattern observed in the average distance data.

D.   Select ONE of the conditions.
     i.   **Design** a controlled experiment to test your explanation for accuracy.
     ii.   **Explain** the probable results of your experiment that would support or disprove your claim.

VI. Practice Test 3   **393**

Section II

2 ▢ Mark for Review

Diabetes mellitus type 2 (T2DM) is a metabolic disorder characterized by insulin resistance on the part of the body's tissues. Onset is usually much later in life and highly associated with obesity.

A.   **Explain** how the citric acid cycle and the electron transport chain are related and how they are affected by exercise.

A recent study evaluated the relative dysfunction of mitochondria of individuals who were lean, obese, and diabetic (type 2). Muscle biopsies were taken both prior to and after a twenty-week exercise program. The mitochondrial mass was measured by cardiolipin, the citric acid cycle was measured by citrate synthase, and electron transport chain activity was measured by NADH oxidase levels. The results are shown below in Table 1.

Table 1. Mean (upper rows) and standard error (lower rows) measurements for markers of ETC activity, in relative units normalized to creatine kinase activity.

Mean	Cardiolipin		Citrate synthase		NADH oxidase	
	Before	After	Before	After	Before	After
Lean ($n = 10$)	70	109	2.4	3.2	0.45	0.63
Obese ($n = 9$)	72	83	3.7	5.2	0.16	0.29
T2DM ($n = 11$)	58	86	3.3	5.1	0.15	0.28

Std Error	Before	After	Before	After	Before	After
Lean ($n = 10$)	6.1	5.2	0.2	0.4	0.1	0.12
Obese ($n = 9$)	7.9	4.8	0.3	0.5	0.05	0.1
T2DM ($n = 11$)	6.5	7.3	0.2	0.5	0.04	0.08

B.   **Construct** a graph of the data indicating citric acid cycle activity. Be sure to use error bars.

C.
  i.   **Identify** which group did not show an increase in cardiolipin after exercise was done.
  ii. **Justify** your choice.

D.   If another measure of the citric acid cycle were added in a future experiment, **explain** what results you would expect.

GO ON TO THE NEXT PAGE.

Section II

---

3 ☐ Mark for Review

Maternal inheritance is one pattern of inheritance which does not follow the rules of Mendelian genetics. It is an example of uniparental inheritance in which all progeny have the genotype and phenotype of the female parent.

Maternal inheritance can be demonstrated in the haploid fungus *Neurospora* by crossing the fungi in such a way that one parent contributes the bulk of the cytoplasm to the progeny. This cytoplasm-contributing parent is called the female parent, even though no true sexual reproduction occurs. The inheritance patterns of a mutant strain of *Neurospora* called poky have been studied using such crosses. Poky differs from the wild-type in that it is slow-growing and has abnormal quantities of cytochromes.

Investigators suspected that the poky mutation was carried in the mitochondria, instead of in the nuclear genome. The following experiments were designed to test this hypothesis.

Step 1: Mitochondria were extracted from poky *Neurospora* mutants.
Step 2: An ultrafine needle and syringe were used to inject these mitochondria into wild-type *Neurospora* cells.
Step 3: These recipient cells were cultured for several generations, and the phenotypes were examined.

Results: The poky phenotype was observed in some of the cultured fungi.

A.   In mitochondrial inheritance the offspring have the genotype and phenotype of the mother. Under Mendelian inheritance it is possible for the offspring to have a different genotype but the same phenotype as the mother. **Describe** how this is possible.

B.   **Explain** what important control(s) the scientists should do to account for the procedure in Step 2.

C.   If the nuclear DNA was injected into the wild-type *Neurospora* cells instead of the mitochondria, **identify** whether the resulting fungi would have the poky phenotype.

D.   **Justify** your prediction.

---

4 ☐ Mark for Review

A diverse flock of finches that can eat many sizes of seeds lives on a small island in Polynesia, which recently experienced a drought. Trees producing robust, medium-sized seeds died out in large numbers, but more resilient trees that produced both small and large-sized seeds are thriving. The finches can shelter in the trees or in burrows underground. The island is home to several tree species that do not produce edible seeds and small rodents that also eat the seeds.

A.   Finches are incapable of digging burrows, yet approximately 25% of the finches found on the island live in underground dwellings. **Describe** the source of the finch dwellings.

B.   **Describe** the relationship between the finches and rodents on the island.

C.   Protein HITB8 is found to be present in high levels in medium-sized beak finches and less so in other finch types. **Predict** what will happen to the frequency of this protein product in the finch population after two generations if the drought continues.

D.   **Justify** your prediction.

GO ON TO THE NEXT PAGE.

VI. Practice Test 3 | **395**

Section II

5 ☐ Mark for Review

Chesapeake Bay is a large estuary on the East Coast of the United States. Freshwater streams and rivers meet the saltwater in the Atlantic Ocean, leading to an area known for rich biodiversity, including waterbirds, fish, crabs, and aquatic vegetation. It is the largest estuary in the United States, stretching across parts of Maryland and Virginia on the East Coast. A food web of some of the native species is shown in Figure 1.

Figure 1. Chesapeake Bay Waterbird Food Web

A.   **Describe** the importance of being the sole occupant of a niche in an ecosystem.

B.   **Describe** the impact of an invasive species that competes with phytoplankton and zooplankton.

C.   **Identify** the number of species with no natural predator on this food web.

D.   **Explain** why toxins are more of a concern for tertiary consumers than they are for primary consumers.

GO ON TO THE NEXT PAGE.

6 ⬚ Mark for Review

A teacher observes squirrels frequently crossing the road and wants to determine if squirrels prefer one side of the road to the other. The north side has predominantly pine trees, while the south side has mostly red maple trees. At 4:00 P.M. for three non-consecutive days the teacher counts the number of squirrels on each side; the data is shown in Table 1.

Table 1

	North Side	South Side
Monday	21	15
Wednesday	18	18
Friday	22	14

A. **Identify** the null hypothesis of this experiment.

B. **Complete** the table with the average data and values necessary to test the hypothesis. (All white empty blanks should be filled.)

	O	E	O-E	$(O-E)^2$	$(O-E)^2/E$
North Side					
South Side					
				$\chi^2 =$	

C. Use statistics to **determine** whether the null hypothesis should be rejected.

D. **Explain** how the results of the experiment support the theory that animal behavior changes as their environment changes.

STOP

END OF EXAM

Practice Test 3: Answers and Explanations

PRACTICE TEST 3 ANSWER KEY

1.	A	21.	B	41.	A
2.	A	22.	C	42.	C
3.	B	23.	C	43.	B
4.	B	24.	B	44.	C
5.	C	25.	D	45.	B
6.	B	26.	A	46.	A
7.	D	27.	B	47.	D
8.	A	28.	B	48.	A
9.	C	29.	B	49.	D
10.	D	30.	A	50.	D
11.	A	31.	D	51.	B
12.	A	32.	D	52.	D
13.	A	33.	D	53.	C
14.	C	34.	B	54.	A
15.	D	35.	A	55.	C
16.	A	36.	C	56.	C
17.	C	37.	D	57.	D
18.	B	38.	A	58.	D
19.	B	39.	A	59.	C
20.	D	40.	D	60.	B

PRACTICE TEST 3 EXPLANATIONS

Section I: Multiple Choice

1. **A** Choice (A) could help remove the molecules that are causing acid rain from the atmosphere. Choice (B) will make acid rain worse, as it will release the molecules higher into the atmosphere. While (C) and (D) may cause harmful environmental effects, they do not contribute to acid precipitation.

2. **A** The Krebs cycle releases carbon dioxide as the carbon molecules are broken down and electron carriers are generated. Glucose is at the start of glycolysis, and pyruvate is at the end of glycolysis. Lactic acid is generated through fermentation.

3. **B** A heterotroph obtains its energy from organic molecules. An autotroph obtains energy from sunlight utilizing pigments such as chlorophyll and uses CO_2 and water to make organic molecules. Therefore, (A) and (C) can be eliminated. Heterotrophs can obtain their energy from ingesting autotrophs, but they can also consume other heterotrophs. Eliminate (D), leaving (B) as your answer.

4. **B** In meiosis, the sister chromatids separate during the second metaphase of meiosis (meiosis II), whereas the sister chromatids separate during metaphase of mitosis. Choice (A) is incorrect because in meiosis there are two rounds of cell division, whereas in mitosis there is only one round of cell division. Chromosomes are replicated during interphase in both meiosis and mitosis, so (C) is incorrect. Choice (D) can also be eliminated because spindle fibers never form during interphase.

5. **C** Amino acids are organic molecules that contain carbon, hydrogen, oxygen, and nitrogen, so eliminate (A), (B), and (D). Don't forget to associate amino acids with nitrogen because of the amino group (NH_2).

6. **B** Unlike eukaryotes, prokaryotes (which include bacteria) do not contain membrane-bound organelles. Bacteria contain circular double-stranded DNA, ribosomes, and a cell wall, so (A) and (D) are incorrect. Also eliminate (C) because bacterial cell membranes are made up of a bilipid layer with proteins interspersed.

7. **D** Populations can be described as the evolutionary unit because changes in the genetic makeup of populations can be measured over time. Eliminate (A), as genetic changes occur only at the individual level. Only under Hardy-Weinberg equilibrium does the gene pool remain fixed over time in a population. However, this statement does not explain why the population is the evolving unit, so (B) is incorrect. Choice (C) is true but does not address the question.

8. **A** Positive feedback occurs when a stimulus causes an increased response. Choices (B), (C), and (D) are examples of negative feedback.

9. **C** A base pair contains two complementary nucleotides, one purine and one pyrimidine. Therefore, there are twice as many nucleotides as there are base pairs.

10. **D** The mean weight of the offspring in the next generation will be heavier than the mean weight of the original population because all the lighter horses in the original population died off. The normal

AP Biology Premium Prep

distribution for weight will therefore shift to the heavier end (to the right of the graph). You can therefore eliminate (C) because the mean weight should increase. The mean weight of the offspring could be heavier or lighter than their parents, so you can also eliminate (A) and (B).

11. **A** Apoptosis (programmed cell death) separates fingers and toes. Meiosis is the formation of gametes, plasmids are circular genetic structures in bacteria, and operons are sets of multiple genes regulated by a single regulatory unit in bacteria.

12. **A** The primitive atmosphere lacked oxygen (O_2). It contained methane (CH_4), ammonia (NH_3), and carbon dioxide (CO_2).

13. **A** Villi and microvilli are fingerlike projections present in the small intestine, which dramatically increase the surface area available for nutrient absorption.

14. **C** The pancreas produces digestive enzymes, and enzymes are sensitive to pH. The inability to neutralize stomach acid would disrupt the function of the enzymes. Food moving too quickly would decrease digestion/absorption rates, but the function of the enzyme would be intact. Removal of water should not greatly affect enzyme function.

15. **D** Ribosomes are the site of protein synthesis. Therefore, the correct answer should start with ribosome, so eliminate (A) and (C). The polypeptide then moves through the rough ER to the Golgi apparatus, where it is modified and packaged into a vesicle. The vesicle then floats to the plasma membrane and is secreted. Choice (D) is the correct answer.

16. **A** Adenine always pairs with thymine, so if 30% of the nucleotides are adenine, then another 30% are thymine. This leaves 40% for guanine and cytosine, which should appear in equal amounts. Therefore, 20% of the nucleotides should be cytosine.

17. **C** A transformation is the uptake of DNA by a bacterium. DNA uptake by a virus, (A), would be a transduction, although this requires a virus to gain the DNA via infection of a bacterium.

18. **B** Some mutations are harmful and decrease the productive success of the individual, so (B) is correct. Dominant alleles would make mutations seem more frequent, as would neutral phenotypic changes, so eliminate (A) and (C). Mutations produce variability in a gene pool. Any gene is bound to mutate. This produces a constant input of new genetic information into a gene pool. Choice (D) is therefore incorrect.

19. **B** The independent variable is the condition the scientist sets up. The dependent variable is the thing that is the measured outcome of the experiment. The scientist should set different temperatures and then measure the germination rate. The outdoor temperature is interesting, but it is too variable to be a good control for the groups with a set temperature.

20. **D** Sympatric speciation occurs when there is no geographic barrier separating the two species (as is the case here). Allopatric speciation requires a geographic barrier.

21. **B** Every cell in an organism has the same set of genes, so eliminate (A). Differences in the timing and expression levels of these genes lead to structural and functional differences, making (B) the correct answer.

402 | For more free content, visit PrincetonReview.com

22. **C** Deoxygenated blood flows through the chambers and vessels labeled 1, 2, 3, and 4 before going to the lungs.

23. **C** The action potentials cause the heart to contract. If chamber 3 is full of more blood, then that would indicate that the preceding chamber, chamber 2 (right atrium), had suddenly contracted a larger amount.

24. **B** Normal cells can become cancerous when a virus invades the cell and takes over the replicative machinery. This would occur if the mammalian genome is altered, such as if the viral genome inserted itself and disrupted mammalian gene expression. A pilus forms between two bacteria, so (A) is wrong. Also eliminate (C) because the host's genome is not converted to the viral genome. Choice (D) is incorrect because spores are released by fungi, not viruses.

25. **D** Crossing-over and synapsis occur during meiosis, which produces haploid gametes, not diploid. Separation of homologous chromosomes occurs during meiosis I, so eliminate (B). Separation of sister chromatids does not occur until meiosis II, so eliminate (C).

26. **A** If the scientist is trying to investigate interspecies breeding, the two potential mates must come from different species; eliminate (B) and (D). If the scientist is hoping to investigate breeding, it is typical to choose a male and a female for most animals; eliminate (C).

27. **B** Choices (A) and (C) are incorrect: asexual reproduction involves the production of two new cells with the same number of chromosomes as the parent cell. If the parent cell is diploid, then the daughter cells will be diploid. Progeny cells are able to fuse together to form a zygote without an outside organism, so eliminate (D). The daughter cells are identical to the parent cell, so choose (B).

28. **B** Cohesion, adhesion, and capillary action are special properties of water that are necessary for transpiration to occur. While high heat capacity is a special property of water, it does not have anything to do with transpiration. Water is a hydrophilic, polar molecule.

29. **B** The two species differ in their preference for water volume, so eliminate (A). Species A prefers 80 mL, and Species B prefers 40 mL. The different watering times do not seem to play a role in growth as there is no obvious pattern and all of the times had similar growth, so eliminate (C). Both plants prefer pH 7, so eliminate (D).

30. **A** The top growth conditions for Species B are pH 7, 40 mL, and any time of day. In those conditions, the Species B plants grew the tallest.

31. **D** The plants in the watering time experiment seemed to be around 60 and 20 cm in height. This corresponds to a pH of 7 rather than 4, so eliminate (A) and (C), and a watering volume of 80 mL rather than 40 mL.

32. **D** Choices (A), (B), and (C) would all give interesting information about the experiment, but the only one that would make the results more statistically significant is to increase the number of plants in each group. Two plants showing the trend is not enough to be sure about the results of the experiment.

33. **D** Photoperiodism is a response of plants to the changing length of day or night. It is a way that plants respond to a signal from the environment. Animal circadian rhythms are also a response to

AP Biology Premium Prep

the environment, making the correct answer (D). Viral infection, (A), increased appetite, (B), and meiotic division, (C), are not responses to the environment.

34. **B** The gross primary productivity is the total amount of energy stored, but the plant has energy needs of its own that are fulfilled with cellular respiration. Choice (A) is true, but not relevant to this question. Choice (C) is not true, because heterotrophs must get energy from autotrophs. Choice (D) is not true because most biomass is stored at the producer level.

35. **A** Choice (B) is incorrect: these organisms exhibit the same behavior because they were subjected to the same environmental conditions and similar habitats. This is an example of convergent evolution. These organisms are not genetically similar, so (C) is incorrect; one is an insect, and the other a bird. They are analogous, so (D) is also incorrect. They exhibit the same function but are structurally different.

36. **C** A mutualistic relationship is a relationship among two organisms in which both benefit. Choice (A) describes parasitism, (B) describes commensalism, and (D) is an example of altruistic behavior.

37. **D** Beta cells secrete insulin. Binding of antibodies to the beta cells in the pancreas will halt the production of insulin, which will lead to an increase in blood glucose levels. Therefore, eliminate (A), (B), and (C).

38. **A** Wobble pairing occurs because the anticodons on the tRNA contain nucleotides that don't care which base in the codon pairs to them. This means that each anticodon might have several possible codons, (A). There are more amino acids than nucleotides, but this is not a cause of the wobble pairing, so eliminate (B). Both (C) and (D) are incorrect: the inosine base would be in the mRNA but this doesn't affect wobble pairing, and each codon could have several anticodons.

39. **A** The mRNA is read 5' → 3', so if the mRNA and tRNA are oriented antiparallel, then the 3' of the mRNA is oriented with the 5' side of the tRNA. The wobble pairing is with the 3rd position on the codon, which would be the 5' end of the tRNA. Choice (D) is incorrect because even though tRNA folds up, it is still a linear chain.

40. **D** Enzymes are one type of protein. They act as catalysts to help speed up reactions by lowering the activation energy of the reaction. When exposed to high temperatures, enzymes denature, and their structure changes. The substrate would no longer be able to bind at the active site, which is why the reaction does not proceed.

41. **A** The disks floated to the surface at a faster rate when exposed to higher concentrations of hydrogen peroxide, so eliminate (C) and (D). The disks did not float to the surface at all when exposed to higher temperatures, so eliminate (B).

42. **C** This is an example of mutualism. Both organisms benefit. Choice (A), parasitism, is a type of symbiotic relationship in which one organism benefits and the other is harmed. Choice (B), commensalism, occurs when one organism benefits and the other is unaffected. Choice (D), endosymbiosis, is the idea that some organelles originated as symbiotic prokaryotes that live inside larger cells.

43. **B** This question tests your understanding of what stage of gene expression utilizes RNA polymerase. Since the presence of RNA polymerase is required for transcription, the regulation is pre-transcriptional.

404 | For more free content, visit PrincetonReview.com

44. **C** Trypsin is an enzyme. Enzymes are proteins and organic catalysts that speed up reactions without altering them. They are not consumed in the process. Therefore, you can eliminate (A) and (B). The rate of reaction can be affected by the concentration of the substrate up to a point, so (D) can be eliminated.

45. **B** DNA helicase unwinds the double helix, so (A) is incorrect. Choice (B), which states that DNA ligase seals the discontinuous Okazaki fragments, is true. DNA polymerase, not RNA polymerase, is the enzyme that causes the DNA strands to elongate, so (C) is incorrect. In the presence of DNA polymerase, DNA strands always grow in the 5' to 3' direction as complementary bases attach. Therefore, (D) is also incorrect.

46. **A** Ecosystems high in variation are more resilient than those with low biodiversity, so eliminate (B). Meiosis provides genetic variation, but mitosis should provide almost none, so eliminate (C). Since crossing-over is a source of genetic variation, it can be a source of speciation; therefore, eliminate (D). Choice (A) is the only correct answer.

47. **D** Each turn of the Krebs cycle takes one pyruvate molecule and generates 1 ATP equivalent in the form of GTP, 2 carbon dioxides, and 4 electron carriers (3 NADH and 1 $FADH_2$), so for both pyruvates the total electron carriers that enter the electron transport chain are 6 NADH and 2 $FADH_2$. The NADH generates 15 ATP, and the $FADH_2$ generates 3 ATP, so 15 + 3 = 18. Therefore, choice (D) is correct. The Krebs cycle on its own makes 2 ATP equivalents per 2 pyruvates, so eliminate (A), 15 would only take the NADH into account and not the $FADH_2$, so eliminate (C), and 10 is the total that is not part of this energy production; eliminate (B).

48. **A** The oxygen released during the light reaction comes from splitting water. Review the reaction for photosynthesis. Therefore, water must have originally contained the radioactive oxygen. Carbon dioxide is involved in the dark reaction and produces glucose, so eliminate (B). Glucose is the final product and would not be radioactive unless carbon dioxide was the radioactive material, so (C) is incorrect. Finally, eliminate (D), because nitrogen is not part of photosynthesis.

49. **D** A ligand binding to a receptor and causing a cellular response is signal transduction. In this case, the ligand is the pheromone, and the response is increased expression of transcription factors. All other answers do not make sense in this context.

50. **D** Homologous structures are organisms with the same structure but different functions. The forelegs of an insect and the forelimbs of a dog are not structurally similar; one is an invertebrate, and the other a vertebrate. They do not share a common ancestor. However, both structures are used for movement, so they are analogous structures. All of the other examples are vertebrates that are structurally similar.

51. **B** Speciation occurred in the Galápagos finches as a result of the different environments on the islands. This is an example of divergent evolution. The finches were geographically isolated. Choice (A), convergent evolution, is the evolution of similar structures in distantly related organisms. Choice (C), disruptive selection, is selection that favors both extremes at the expense of the intermediates in a population. Choice (D), stabilizing selection, is selection that favors the intermediates at the expense of the extreme phenotypes in a population.

AP Biology Premium Prep

52. **D** Mutations produce genetic variability. All of the other answer choices are forms of asexual reproduction.

53. **C** Secondary consumers feed on primary consumers. If you set up a pyramid of numbers, you'll see that the herrings belong to the third trophic level.

54. **A** The biomass is the total bulk of a particular living organism. The phytoplankton population has both the largest biomass and the most energy.

55. **C** A decrease in the herring population will lead to an increase in the number of crustaceans and a decrease in the phytoplankton population. Reorder the organisms according to their trophic levels and determine which populations will increase and decrease accordingly.

56. **C** Relatedness is shown on a phylogenetic tree by how far back you have to go to reach the common ancestor of the two species (shown by a branch node). Since the recent species are on the right, traveling to the left goes back in time. Because you have to go to the oldest common ancestor to connect *B. musculus* and *B. brydei*, they are the least closely related. Even though they are near each other on the tree, clades can rotate around a node, so this does not matter.

57. **D** Because molecular biology demonstrates where the actual mutations occurred (rather than just what phenotype is observed), it is the most reliable; it allows you to detect changes that are not phenotypically observable.

58. **D** Because the brain is destroyed, it is not associated with the movement of the leg. Choice (A) is incorrect, as reflex actions are automatic. Choices (B) and (C) can also be eliminated; both of these statements are true but are not supported by the experiment.

59. **C** The DNA template strand is complementary to the mRNA strand. Using the mRNA strand, work backward to establish the sequence of the DNA strand. Don't forget that DNA strands do not contain uracil, so eliminate (B) and (D).

60. **B** Use the amino acid chart to determine the sequence after uracil is deleted. The deletion of uracil creates a frameshift.

Section II: Free Response

Short student-style responses have been provided for each of the questions. These samples indicate an answer that would get full credit, so if you're checking your own response, make sure that the actual answers to each part of the question are similar to your own. The structure surrounding them is less important, although we've modeled it as a way to help organize your own thoughts and to make sure that you actually respond to the entire question.

Note that the rubrics used for scoring periodically change based on the College Board's analysis of the previous year's test takers. We've done our best to approximate their structure, based on our institutional knowledge of how past exams have been scored and on the information released by the test makers.

Our advice is to over-prepare. Find a comfortable structure that works for you, and really make sure that you're providing all of the details required for each question. Also, continue to check the College Board's website, as they may release additional information as the test approaches. For some additional help, especially if you're worried that you're not being objective in scoring your own work, ask a teacher or classmate to help you out. Good luck!

406 | For more free content, visit PrincetonReview.com

AP Biology Premium Prep

Question 1

A. **Describe** the pattern of distribution of the starfish using the provided data. Be specific as to how the data changes based on the time of year. (1 point)

The starfish are closer together during colder months and spread farther apart during warmer months.

- 1 point for **describing the pattern** and **specifying the times of the year** the pattern occurs

B. **Identify** three physical or ecological conditions that could cause the starfish to change their distance from other starfish. (1 point)

Three conditions that could cause the pattern in distance between starfish are temperature, light, and predation.

- 1 point for identifying any three **testable** variables, such as competition, metabolism, salinity, temperature, etc.

C. **Explain** how each condition could bring about the pattern observed in the average distance data. (3 points)

During the winter months in North America, average temperature is lower. Lower temperatures could cause starfish to huddle together to conserve warmth. During the winter months, less light is available. Less light could cause starfish to huddle together in similar locations if their prey is drawn to specific locations due to limited light. During the winter months, fish and turtles that prey on starfish may move to specific locations that receive more light or warmer conditions, which may push the starfish to migrate to a limited number of safe locations.

- 1 point for EACH **clear and reasonable explanation** of the condition and how it could **influence the observed pattern**

D. Select ONE of the conditions.

   i. **Design** a controlled experiment to test your explanation for accuracy. (5 points)

To test if light is causing starfish to huddle closer together in cold months, set up starfish in a large square tank with light bulbs of different strengths aimed at each corner. Hold all other variables constant, including amount and type of water, size of tank, amount and type of food provided to the starfish, species, size, age and gender of the starfish, and conditions in the laboratory. Measure the average inches between the starfish at even intervals. If more light is provided to a corner of the tank, more starfish will migrate to that corner of the tank, thus reducing the average distance between them. Repeat the experiment four times, rotating the light bulbs each time.

- 1 point for EACH of the experiment components: **constants, independent variable, dependent variable, hypothesis, and verification**

VI. Practice Test 3: Answers and Explanations | **407**

AP Biology Premium Prep

   ii. **Explain** the probable results of your experiment that would support or disprove your claim. (2 points)

The starfish are likely to cluster together in corners with more light. Starfish feed on prey that feed on microorganisms attracted to light, so smaller regions of the tank will be suitable for inhabiting at low temperatures. These results support my hypothesis, as the starfish will huddle closer together and result in smaller average inches between starfish.

- 1 point for describing the **probable results** of your experiment
- 1 point for specifying if the results **support or disprove** your hypothesis

Question 2

A. **Explain** how the citric acid cycle and the electron transport chain are related and how they are affected by exercise. (2 points)

The citric acid cycle (Krebs cycle) is an important part of cell respiration. The reduced electron carriers created during the citric acid cycle bring their electrons to the electron transport chain, which is used to generate ATP. During exercise the muscle cells will need more energy. The process of cell respiration will speed up. When muscle cells have exhausted their stores of oxygen, then the electron transport chain will cease and fermentation will take over.

- 1 point for explaining how the **citric acid cycle creates electron carriers which provide electrons to the electron transport chain to make ATP**
- 1 point for explaining how exercise creates a need for more energy, which will **speed up both cycles**; eventually the muscle cells will run out of resources and **switch to fermentation**

B. **Construct** a graph of the data indicating citric acid cycle activity. Be sure to use error bars. (4 points)

Citrate Synthase Activity

- 1 point for accurate graphing of the data (using **citrate synthesis** column)
- 1 point for appropriate axes ranging from approximately **0 to 6** on the y-axis AND the **three types of participants** on the x-axis

408 | For more free content, visit PrincetonReview.com

AP Biology Premium Prep

- 1 point for an appropriate title relating **citric acid cycle activity**, appropriate label of **relative units** on the *y*-axis, appropriate label of **type of participant** on the *x*-axis, AND a key for **before and after**

- 1 point for accurate graphing of error bars

C.

    i. **Identify** which group did not show an increase in cardiolipin after exercise was done. (1 point)

The obese group did not show an increase after exercise.

- 1 point for identifying the **obese group** as not showing an increase in cardiolipin after exercise

    ii. **Justify** your choice. (1 point)

Although the numbers 72 and 83 seem different, when the standard error is considered, the values overlap: $72 + 7.9 = 79.9$ and $83 - 4.8 = 78.2$. The lean and T2DM participants each saw an increase in cardiolipin with exercise.

- 1 point for justifying by citing SPECIFIC evidence from the **graph** AND/OR **Table 1**

D. If another measure of the citric acid cycle were added in a future experiment, **explain** what results you would expect. (2 points)

If another measure were included, there would be an increase for all groups after exercise. If one measure of citric acid cycle increases, such as citrate synthase, then all measures of the citric acid cycle should increase.

- 1 point for predicting an **increase** for all three groups
- 1 point for explaining that the increase will occur because the data already shows an increase in the citric acid cycle before and after; **additional measures should show the same results**

Question 3

A. In mitochondrial inheritance the offspring have the genotype and phenotype of the mother. Under Mendelian inheritance it is possible for the offspring to have a different genotype but the same phenotype as the mother. **Describe** how this is possible. (1 point)

It is possible for a mother and offspring to have a different genotype and the same phenotype under classic Mendelian inheritance because there are more than one genotype that have the same phenotype. For example, a homozygous dominant and a heterozygous genotype will both have the dominant phenotype.

- 1 point for describing how Mendelian inheritance has **multiple genotypes that have the same phenotype**

B. **Explain** what important control(s) the scientists should do to account for the procedure in Step 2. (1 point)

It would be important to make sure the injection process itself is not causing a problem. There should be a control group of wild-type cells that is injected with saline to account for the injection and make sure it is the actual mitochondria that is causing the mutant phenotype.

- 1 point for explaining how a **control group should be injected with a saline solution** to ensure **the injection itself is not causing the results**

VI. Practice Test 3: Answers and Explanations | **409**

AP Biology Premium Prep

C. If the nuclear DNA was injected into the wild-type *Neurospora* cells instead of the mitochondria, **explain** whether the resulting fungi would have the poky phenotype. (1 point)

They should NOT have the poky phenotype.

- 1 point for identifying that the fungi should **NOT** have the poky phenotype

D. **Justify** your prediction. (1 point)

The resulting fungi should not have the poky phenotype because the trait is not carried on a nuclear genome. It is part of the mitochondrial genome, and giving the nuclear genome is not going to provide the gene for the poky phenotype.

- 1 point for justifying that **the gene is on the mitochondrial genome, not the nuclear genome**, so injecting it into normal cells wouldn't cause any change

Question 4

A. Finches are incapable of digging burrows, yet approximately 25% of the finches found on the island live in underground dwellings. **Describe** the source of the finch dwellings. (1 point)

The source of the burrows must come from some other animal on the island. Small rodents could dig the burrows which the finches live in opportunistically, either simultaneously or after the rodents have abandoned the burrows.

- 1 point for describing **another organism that could create the underground dwellings**, such as a rodent

B. **Describe** the relationship between the finches and rodents on the island. (1 point)

The finches and the rodents have a symbiotic relationship. The rodents may provide shelters for the birds, and the birds spread seeds which will eventually develop into a food source for the rodents as well. This describes a mutualistic behavior. Otherwise, the relationship could be commensalistic, where the birds benefit but the rodents do not benefit from the birds.

- 1 point for describing a **symbiotic** relationship between the finches and the rodents

C. Protein HITB8 is found to be present in high levels in medium-sized beak finches and less so in other finch types. **Predict** what will happen to the frequency of this protein product in the finch population after two generations if the drought continues. (1 point)

The protein HITB8 will decrease in abundance.

- 1 point for predicting that the frequency of protein HITB8 will **decrease**

D. **Justify** your answer. (1 point)

A drought will make surviving on the island more difficult due to the birds' decreased ability to find the food with their beak type. As such, the frequency of the gene encoding the HITB8 protein will be less plentiful, and the protein will be expressed in lower numbers.

- 1 point for justifying by describing how the drought will make some of the finches die out, as they **rely on trees that produce medium-sized seeds**, and **the drought is making those trees die out**

410 | For more free content, visit PrincetonReview.com

Question 5

A. **Describe** the importance of being the sole occupant of a niche in an ecosystem. (1 point)

A niche is helpful because it is a set of resources that an organism is well suited to utilize. Having no other competitors in the niche means that there is a set of resources (physical space, food, water, etc.) that the organism will have exclusive access to.

- 1 point for describing that being the sole occupant means there will be **little to no competition for necessary resources**, making it much easier to survive

B. **Describe** the impact of an invasive species that competes with phytoplankton and zooplankton. (1 point)

If an invasive species competed with phytoplankton and zooplankton they would possibly disappear. Invasive species often have no natural enemies and they can steal resources from the other occupants in their niche that still do have to worry about predators. The consumers of the zooplankton and phytoplankton might also disappear when their food source disappears and an invasive species can cause trouble throughout the food web.

- 1 point for describing how **an invasive species will likely not have any predators**, so it will be able to steal resources away from the phytoplankton and zooplankton, which will eventually **impact the herbivores and primary consumers that rely on them and have effects throughout the food web**

C. **Identify** the number of species with no natural predator on this food web. (1 point)

There are 7 species that have no natural predator (they have no lines pointing away from them).

- 1 point for identifying the number of species as **7**

D. **Explain** why toxins are more of a concern for tertiary consumers than they are for primary consumers. (1 point)

Toxins are a concern higher up the food web because they can often become concentrated. For example, it is bad for a toxin to be absorbed by a plant, but then a fish eats 10 toxin-affected plants and a bigger fish eats 10 of the little fish, which is like 100 toxin-affected plants. Higher up the food web toxins can concentrate, and the animals can begin to be harmed by the level of toxin even though the primary consumers were not.

- 1 point for explaining how **toxins become more concentrated up the food web**, as **higher-level consumers eat larger quantities of lower-level consumers and producers** who may have the toxin in their cells

AP Biology Premium Prep

Question 6

A. **Identify** the null hypothesis of this experiment. (1 point)

Squirrels do not prefer one side of the road over another.

- 1 point for correctly identifying the null hypothesis as the fact that **squirrels do not have a preference for one side of the road over the other**

B. **Complete** the table with the average data and values necessary to test the hypothesis. (All white empty blanks should be filled.) (1 point)

	O	E	O–E	$(O–E)^2$	$(O–E)^2/E$
North Side	20.33	18	2.33	5.4289	0.3016
South Side	15.66	18	–2.33	5.4289	0.3016
				$\chi^2 =$	0.603211

- 1 point for completing all white blanks of the table; the table is based on the chi-square equation found on the Equations and Formulas sheet

C. Use statistics to **determine** whether the null hypothesis should be rejected. (1 point)

Since our degree of freedom is 1, and we are using $p = 0.05$, our critical value is 3.841. Our χ^2 value is less than our critical value, so we fail to reject our null hypothesis.

- 1 point for determining that the null hypothesis should **NOT be rejected** because the calculated chi-square value is **smaller** than the chi-square critical value (on the Equations and Formulas sheet)

D. **Explain** how the results of the experiment support the theory that animal behavior changes as their environment changes. (1 point)

According to our data, the squirrels do not prefer one side of the road over the other. As the road was constructed, what was likely a continuous forest became separated. As both pine trees and maples provide food and shelter resources, the squirrels have adapted to the change in their habitat by migrating to both sides of the road frequently.

- 1 point for explaining how the **results link to changes in the environment and the squirrel behavior**

AP Biology Premium Prep

HOW TO SCORE PRACTICE TEST 3

Section I: Multiple Choice

_____ × 0.57 = _____
Number Correct   Weighted
(out of 60)   Section I Score
                              (Do not round)

Section II: Free Response

Question 1:

_____ × 1 = _____
(out of 9)   (Do not round)

Question 2:

_____ × 1 = _____
(out of 9)   (Do not round)

Question 3:

_____ × 1 = _____
(out of 4)   (Do not round)

Question 4:

_____ × 1 = _____
(out of 4)   (Do not round)

Question 5:

_____ × 1 = _____
(out of 4)   (Do not round)

Question 6:

_____ × 1 = _____
(out of 4)   (Do not round)

Sum =

_____
Weighted
Section II Score
(Do not round)

AP Score Conversion Chart Biology	
Composite Score Range	AP Score
57–68	5
45–56	4
36–44	3
21–35	2
0–20	1

Composite Score

_____ + _____ = _____
Weighted   Weighted   Composite Score
Section I Score   Section II Score   (Round to nearest
                                            whole number)

VI. Practice Test 3: Answers and Explanations | **413**